The Misuse, Misrepresentation, and Politicization of Statistics in American Society

The Misuse, Misrepresentation, and Politicization of Statistics in American Society

Robert E. Parker

LEXINGTON BOOKS
Lanham • Boulder • New York • London

Published by Lexington Books
An imprint of The Rowman & Littlefield Publishing Group, Inc.
4501 Forbes Boulevard, Suite 200, Lanham, Maryland 20706
www.rowman.com

86-90 Paul Street, London EC2A 4NE

British Library Cataloguing in Publication Information Available

Library of Congress Cataloging-in-Publication Data

Names: Parker, Robert E., 1957- author.
Title: The misuse, misrepresentation, and politicization of statistics in American society /
    Robert E. Parker.
Description: Lanham : Lexington Books, [2022] | Includes bibliographical references
    and index.
Identifiers: LCCN 2021057322 (print) | LCCN 202105323 (ebook) |
    ISBN 9781793625526 (cloth) | ISBN 9781793625540 (paper) | ISBN
    9781793625533 (ebook)
Subjects: LCSH: Statistics—Political aspects—United States. | Vital statistics—
    United States.
Classification: LCC HA179 .P37 2022 (print) | LCC HA179 (ebook) | DDC
    304.60973–dc23/eng/20211206
LC record available at https://lccn.loc.gov/2021057322
LC record available at https://lccn.loc/gov/2021057323

This book is dedicated to my wife, Vianett Achaval, our daughter Penelope, and to the memory of our canine companion, Patience.

Contents

Acknowledgments   3

Introduction   1

Chapter One: The BLS and the Underestimation of Unemployment   11

Chapter Two: The Centers for Disease Control and the
    Overestimation of Life Expectancy   33

Chapter Three: The FBI and the Miscalculation of Crime   53

Chapter Four: The Census Bureau and the Decennial Population
    Undercount   69

Chapter Five: Why Official Statistics Matter   87

References   101

Index   123

About the Author   135

Acknowledgments

I would like to acknowledge my editor, Courtney Morales, for expressing interest in my work and for guiding the manuscript through to completion, and to Foster Kamanga for the generous support he provided in creating time for me to focus on completing the book.

Introduction

THE SOCIAL SIGNIFICANCE OF OFFICIAL STATISTICS

Americans are increasingly inundated with official statistical information about significant social issues. The growing number of authoritative reports warrants sustained scrutiny as their economic and political implications multiply. This book critically examines key official statistics created by four agencies—the Bureau of Labor Statistics (BLS), the Centers for Disease Control and Prevention (CDC), the Federal Bureau of Investigation (FBI), and the Bureau of the Census. In the following pages, the offices that create the unemployment rate, life expectancy statistics, national crime reports, and national demographic figures will be examined. Particular attention is paid to the way official statistics have been politicized to reaffirm a society increasingly characterized by the prevailing neoliberal ideology. After surveying the origins of these four government agencies, the book outlines their economic and political significance. In so doing, readers will gain a better comprehension of why official government statistics matter and why they deserve sustained critical scrutiny.

The official statistics covered here have critical implications both for social policies and personal decision-making. For example, the official estimates of the population derived from the decennial census determine the distribution of elected representatives and the allocation of two trillion dollars in local and state funding for social and physical infrastructure projects. Another illustration of the social significance of official statistics is the way the CDC's "official" estimates of ever-increasing longevity among Americans has been used to justify raising the full retirement age (FRA) for Social Security beneficiaries from sixty-two to sixty-seven years. As with these, the other official statistics covered in this manuscript are discussed within the context of a broader social milieux that makes their political implications more transparent.

1

OVERVIEW OF CHAPTERS

After the introductory chapter surveys the book's major themes, the four chapters that follow comprise the substantive portion, outlining many of the key controversies surrounding each of the statistical series.

Chapter 1

Chapter 1 provides a critical examination of the Bureau of Labor Statistics (BLS) and its monthly estimation of employment, unemployment, and the gray areas in between work and joblessness. The modern day BLS is modeled on early state bureaus of labor statistics that were first organized in the latter half of the nineteenth century. This chapter details early efforts to measure unemployment and the way state agencies paved the way for the creation of the federal Bureau of Labor Statistics. Importantly, the first chapter provides a broader social context in which to view the remainder of the book. The transition from state bureaus of labor statistics to a national agency reveals the social and historical evolution that government agencies sometimes undergo, and how their initial missions can be undermined.

After providing historical background on the BLS, the first chapter focuses on one of the most misused, misinterpreted and politicized social statistics disseminated from any official source—the monthly unemployment rate. Although the official unemployment rate was 5.2 percent in August 2021, other statistical parameters (showing broader measures of labor force utilization) reveal that the true unemployment rate has a long history of being underestimated. Some of these supplemental labor force metrics include categories previously used by the Bureau, including discouraged workers and those experiencing long-term joblessness. The politicization of the monthly unemployment rate is of particular significance among those holding elected power, and among those seeking formal political office. To illustrate, presidential candidate Jimmy Carter effectively used a so-called "misery index" (the unemployment rate combined with the inflation rate) to help defeat Gerald Ford in 1976. But in 1980, Carter found his fortunes reversed when Ronald Reagan used the same, now worsening index to help doom the incumbent's chances for a second term (Nutting 2015).

Chapter 2

Chapter 2 examines another official statistic with tangible social policy implications—official life expectancy. In this chapter, the Center for Disease Control's (CDC) rapid rise from a modest agency housed in a single floor of an office building in downtown Atlanta to a full-fledged transnational organization undertaking sweeping obligations in more than fifty nations is chronicled. This chapter reveals that the official life expectancy data published by the CDC exaggerates the extent to which Americans are experiencing greater longevity. The widely circulated official reports regarding increasing life expectancy has effectively enabled conservative political interests who have successfully used the data to lobby for raising the full retirement age for Social Security beneficiaries from sixty-two to sixty-seven years of age.

Chapter 3

Following the chapter on the CDC and the ramifications of its official life expectancy statistics, the creation and implications of official crime figures are explored. When the FBI (through the Bureau of Justice Statistics) releases its annual Crime in the United States report each fall, few are aware that it is based on a relatively small number of "street" crimes that the law enforcement agency deems most worthy of tracking. Far from being a comprehensive survey of crime, Part I of the Bureau's Uniform Crime Report includes murder and nonnegligent manslaughter, forcible rape, robbery, aggravated assault, burglary, larceny-theft, motor vehicle theft, and arson.

From nearly 20,000 law enforcement agencies, the FBI gathers data on twenty-one additional unlawful acts, but it is the Part I index that the mass media highlights as the nation's "official" crime rate. In its Part II Crime Index, the FBI includes many white-collar transgressions that are far more injurious to society financially, including forgery, counterfeiting, fraud, and embezzlement. But these misdeeds seldom garner the same attention as Part I offenses. Moreover, the FBI's crime index has significant methodological issues that undermine its continued use as an accurate snapshot of crime in the nation. For example, the nearly 20,000 law enforcement agencies from which the FBI gathers information are not required to submit reports, they do so voluntarily. Moreover, variations among law enforcement agencies in the way data are collected has historically led to gross errors in the reporting of certain types of crime. In 1973, the National Crime Victimization Survey (NCVS), which gathers information on individuals and households, was introduced to address some of the major measurement issues involved with the Uniform Crime Report (UCR). The differences between the UCR and NCVS will be more fully reviewed in chapter 3.

Chapter 4

Here, the book turns to the Census Bureau, its social and political history, and the most significant statistic it produces—its estimate of the US population—drawn from its decennial canvassing. Efforts to count the population were undertaken even before the national census began gathering official demographic data in 1790. That first census, and every one since, has been used to apportion members to the US House of Representatives and to distribute what now amounts to trillions of dollars in federal funding for social and physical infrastructure programs. Like other official statistical series, there have been fundamental controversies surrounding the census since its inception.

Specifically, there have been persistent concerns about who the Bureau "counts" and whether it can ever fully enumerate the US population. Not counting, or undercounting marginalized groups, has been a particularly noteworthy issue since the Bureau's earliest days. Today, being low income, black, Hispanic, or Native-American translates into a greater chance of being discounted when it comes to political representation and federal funding. Minority groups, particularly blacks, have long been disadvantaged by the decennial undercount. Because the census overcounts whites and undercounts people of color, social programs that should be allocated to minority communities are misdirected, thereby reinforcing systemic racism.

Chapter 5 and Major Themes

The final chapter ties together the major social, economic and political themes found throughout the book. Emphasis is paid to the commonalities as well as the differences in the way official statistics have been used, misused, misinterpreted, and politicized. For instance, in one fashion or another, all of the statistics under examination here have been politicized by powerful interests seeking to advance their national policy agenda. Another common focus is the way government agencies create conceptual categories that determine who counts, and who does not count in society. For instance, someone who is merely unemployed does not "officially" count according to the BLS, but being without work and making an active, demonstrable effort to find work does satisfy the Bureau's criteria for being labeled "unemployed."

Finally, a major theme of the book is that official statistics serve to frame economic and social reality for consumers of the data. Official statistics are intricately integrated into mainstream media coverage in the United States. But they also serve as a media in their own right, influencing Americans to concentrate their attention in particular directions, while ignoring, or minimizing others. In this way, official statistics can easily be misinterpreted. As the ripple effects of the pandemic spread in 2020, individuals who relied entirely on "official" government versions of reality would have had an unrealistically low estimation of the actual number of Americans dying from COVID-19, an inaccurately low assessment of the number of individuals who became unemployed during the recession, a distorted view of the number of minorities in the country, and little insight, into the growing problem of domestic violence.

Lastly, the final chapter provides an opportunity to consider other government agencies and the statistical series they produce. Specifically, it looks at the Commerce Department's Gross Domestic Product (GDP) and Bhutan's alternative index of well-being—the Gross National Happiness Index. As with the other statistical series considered here, a major concern with the GDP surrounds the issue of what the metric counts and what it excludes. The GDP has been widely criticized for emphasizing formal economic transactions, while neglecting the underground economy. In short, the GDP does not measure all economic activity, only those that are legally sanctioned. As presently constructed, the GDP leaves out large sectors of the subterranean economy such as illicit drug sales and prostitution, that if included, would add to the nation's chief measure of national well-being.

In sum, this book maintains that official statistics matter. They matter because they have direct effects on the lives and livelihoods of US residents. They matter because they drive the federal government's fiscal and monetary policies and determine how funding for programs at the state level will be allocated. They matter because they have an impact on the decisions people make about their daily lives. Official statistics warrant critical scrutiny because they tell us who matters in society . . . who counts as being unemployed, who counts as a criminal, and who counts as officially residing in the country. Critically, official statistics matter because they shape our perceptions about the health of the economy and society as a whole. Official statistics on public health, the extent of unemployment, the amount of crime, and the US population paint a consistently positive, but distorted image of the nation.

WHY OFFICIAL STATISTICS MATTER: COUNTING COVID'S FATALITIES

The outbreak of the coronavirus pandemic demonstrated as well as any event in recent history why official statistics matter. Developments surrounding the pandemic, particularly allegations of under and overcounts, illustrated how official statistics are misused. It also shows how their interpretation can be influenced by political forces. The experience of the pandemic illustrated, from the highest levels of institutionalized power, how official statistics are routinely politicized by public figures to advance their own special interests. Finally, the publication and dissemination of official data on the virus showed how official statistics shape our perceptions of everyday life. The way

Americans spent their days during periods of enforced mitigation measures depended in part on the extent to which they believed the CDC's numbers were accurate. In the following discussion, some of the ways official health statistics were misused, misinterpreted and politicized will be reviewed.

By the time most had became aware of the coronavirus outbreak, examples of statistical illiteracy abounded. The mainstream media routinely compared the number of deaths from the coronavirus outbreak to deaths from wars in which the United States has been involved, or from violent attacks against the nation. Accounts of how the pandemic killed more people than had died on the day Pearl Harbor was attacked, or on 9/11, or from wars in Korea, Afghanistan, or Vietnam were frequently featured in headlines. For example, O'Connor stated that "the number of people killed by the novel Coronavirus in the United States has exceeded US casualties incurred throughout the conflict in Afghanistan, the longest war in US history" (2020). The Hill proclaimed "the coronavirus death toll in New York City this week exceeded the number of people killed in the World Trade Center on September 11, 2001" (2020). And National Public Radio reported that in less than three months, more lives had been lost to the pandemic than the 58,000 Americans who died in Vietnam over two decades (Welna 2020). Other statistically inappropriate comparisons included deaths attributable to natural disasters, such as Hurricane Katrina.

Rather than making meaningful contrasts, the mass media used "apples to oranges" comparisons that made for sensation-grabbing headlines. Few print or broadcast outlets attempted to engage their audiences about total fatalities attributable to the seasonal flu, or dug deeply into the number of lives lost to other communicable diseases. An exception was *The Business Insider* which published an article featuring an informative chart documenting that COVID-19 had killed more than Sars, Mers or Ebola, but fewer than the swine flu (Brueck and Gal 2020). The contentious debate over the official death count sowed the seeds of anxiety among the public and served to mobilize anger toward government-imposed lockdowns associated with the public health crisis. As this case demonstrates, the veracity of official statistics, or the lack thereof, can play an important role in maintaining confidence in the government and its programs.

Politicizing Covid's Casualties

As the coronavirus claimed a growing number of American lives, calculating the official number of deaths, and politicizing the "official" numbers became commonplace. No less a political figure than the president of the United States publicly expressed doubts about the virus and its lethality. In April of 2020, with much of the nation in a lockdown, President Trump suggested that New York City was artificially inflating its death toll after a revised count added more than 3,700 fatalities to the city's tally. Trump's accusation was met with expressions of bewilderment from the mayor's and governor's offices (Siverstein 2020). Later, the president complained to his advisers about the way coronavirus deaths were being calculated, implying the number of deaths was substantially lower. Further, a report by *Axios* revealed that a significant number of Trump's senior aids shared his view about an alleged death overcount (2020).

As time passed, the president continued minimizing the severity of the pandemic, often in direct contradiction to the White House Coronavirus Task Force, headed by Vice President Pence. During the summer, the president attempted to focus Americans' attention on reopening the economy rather than on public health measures (even those his administration had ordered). According to recent research, the president's efforts had palpable consequences, influencing the way fatalities were being counted in different parts of the country. According to *Stat* "tens of thousands of COVID-19 deaths are going unreported in the U.S., with far more missed in counties that strongly supported former President Trump" (2021).

Conservative media pundits consistently contended that the number of fatalities attributable to the coronavirus were overstated (Lewis 2020; Blake 2000). Commentators who argued that the death count was being overreported included Candace Owens. Owens maintained that the pandemic's toll was being manipulated in order to gain a political advantage for the Democrats in the fall 2020 national election. In a widely cited tweet, Owens wrote, "Apparently, doctors and nurses around the world are wondering why no one is dying from heart attacks or strokes any more . . . flu and pneumonia deaths also went off a cliff. Turns out everyone is only dying of coronavirus now. Gee, I wonder why" (Greenberg 2020).

Likewise, Fox News' Britt Hume tweeted that New York's fatality numbers were being overreported. He then appeared on Tucker Carlson's late-night talk show claiming that any person with the virus who died was being counted as a COVID-19 death regardless of other health conditions (Ecarma 2020). He added there may be "reasons people seek an inaccurate death count" (Darcy 2020). The late conservative commentator Rush Limbaugh, who received the presidential Medal of Freedom from President Trump, dismissed the coronavirus as being similar to the common cold. Limbaugh stated "it's admittedly speculation, but at this point, what if we are recording a bunch of deaths to coronavirus which really should not be chalked up to coronavirus?" "People die on this planet every day from a wide variety of things" (Baragona 2020).

Anthony Fauci, the Director of the National Institute of Allergy and Infectious Diseases, attributed the allegations to conspiracy theories swirling around the outbreak of the coronavirus. According to Fauci, "any time we have a crisis of any sort there is always this popping up of conspiracy theories" (Milman 2020). Many medical practitioners joined Fauci in highlighting the role of conspiracy theories in undermining the efforts of health care workers to ameliorate the virus's impact. More than a few were deeply offended by President Trump's statement that "doctors get more money and hospitals get more money" if they say people died from COVID-19 rather than their comorbidity" (Dwyer 2020).

Throughout the first year of the pandemic, the prevailing view among most public health experts, politicians and media commentators was that the death toll from the coronavirus was being understated. New York City councilman Mark Levine stated there was "no doubt the official death toll was an undercount" (Abdelmalek et al. 2020a). Later that month, an ABC Report cited public health experts who claimed the death count was understated by tens of thousands (Abdelmalek et al. 2020b). Likewise, *The Philadelphia Inquirer* asserted that thousands of deaths from the virus were being uncounted in Pennsylvania and New Jersey alone (Williams and Avril 2020).

By summer, Reuters was reporting that the number of Americans who died from the coronavirus between March and May was likely significantly higher than the official US count. According to their report, the undercount was due in part to state-level reporting discrepancies" (Beasley 2020). Later in the year, the prestigious medical journal *Lancet* published a study that had profound implications for the debate over the official count. The study found that during the first wave of the pandemic, fewer than 10 percent of the US adult population formed antibodies against the virus, and fewer than 10 percent of those with antibodies were diagnosed, signaling a substantial undercount of total cases in the nation (Anand et al. 2020). An important indicator that the official death toll was being understated was the extent of "excess mortality." In early 2021, a study from Boston University found that between the beginning of February 2020 to mid-October 2020, there were more than 88,000 "excess" deaths in 787 US counties that were not a part of the official coronavirus death toll (Stokes et al. 2021).

The sense that deaths due to the coronavirus were not being fully reported was not limited to the United States. And neither was the politicization of statistics surrounding the outbreak. Worldwide, skeptics raised doubts that governments were accurately reporting deaths. In the United States, government intelligence agents expressed the belief that China vastly underreported its death toll throughout the early stages of the pandemic. Their investigation suggested China concealed the extent of the coronavirus outbreak, understating both total cases and the number of deaths (Wadhams and Jacob 2020). In Brazil, the data published about coronavirus cases and deaths was sharply curtailed as President Jair Bolsonaro grew increasingly uncomfortable with the country's status as one of the world's most severe pandemic hotspots

(Leite et al. 2020). Late in 2020, the Russian government conceded that they had been underestimating deaths from the coronavirus. Tatiana Golikova, the country's prime minister, stated that Russia's true death toll was not 57,000, as official figures claim, but more than 180,000 (Palash 2021). At some point in 2020, virtually every nation's death total was being critiqued, particularly by those working on the front lines of the crisis (Chatterjee 2020).

SUMMARY: THE COVID-19 PANDEMIC AND WHY OFFICIAL STATISTICS MATTER

The recent coronavirus pandemic has provided numerous illustrations of the ways official statistics are misused, misinterpreted and politicized. As noted earlier, since the outbreak of the coronavirus pandemic, abundant charges have been leveled by skeptics that the number of covid-related fatalities were being overcounted by official government agencies. Counterclaims that the official figures amounted to an undercount were equally plentiful. Additionally, the pandemic ushered in a new wave of critics of official unemployment numbers (as state unemployment insurance systems crashed under the weight of an unprecedented wave of layoffs). Finally, the 2020 decennial census was complicated by the pandemic as it interfered with the ability of the Bureau's enumerators to locate and record "hard-to-count" groups for the decennial census.

In short, the recent experience with the COVID-19 pandemic demonstrates that official statistics matter because when social data are inaccurate or distorted, it generates a ripple effect of social, economic, and political implications. For example, if the number of Americans who were dying from the covid pandemic were being overstated, it would expedite the ability of government authorities to adopt public health policies that were unnecessarily restrictive, inimical to the economy as a whole, and specifically harmful to workers displaced by the virus-induced recession. In contrast, if the CDC's numbers were being undercounted, it would lead to a misallocation of public (and private) health resources available to combat the contagion. Further, emergency measures that potentially could have been drawn upon to tackle the public health crisis in a more aggressive manner would likely go unconsidered if official numbers consistently understated the magnitude of the problem. In short, whether the official numbers were inaccurate because they were understating the problem, or the obverse, they matter because they impact the integrity of the nation's statistical reporting system. To the extent that the official numbers are routinely criticized, it serves to undermine the trust and confidence citizens have in the government to accurately record and appropriately address social problems. In short, the covid pandemic high-lighted the social significance of taking a serious, sustained, and scholarly look at official statistics and the agencies that produce them.

Chapter One

The BLS and the Underestimation of Unemployment

The social significance of work today cannot be overestimated. After the family, no social institution is more central to people's lives than what they do to earn a living. For most, one's occupation becomes their primary means of social identification. Academic studies reveal a great deal of variation among workers regarding the sense to which they find jobs satisfying. Research also suggests many workers today are engaged in routine work that is socially counterproductive. Yet for the majority, even a weak attachment to a boring job is a more acceptable alternative to being without work. Few issues can be more detrimental to one's self-esteem than being unemployed. The emphasis in society on individualism plays a major role in creating feelings of anxiety among those experiencing unemployment. Despite the fact that the economy has never generated enough jobs for everyone, displaced workers often feel being out of work is a personal problem.The neoliberal ideology guiding social policy maintains there are enough jobs for those in the labor force and that the unemployed are simply victims of their own lack of initiative and an overly-generous welfare system.

The capacity to make meaningful contributions to family and society become increasingly problematic for the unemployed, especially among the long-term jobless. As Herbert Gans noted decades ago,

> Unemployment may be the most dangerous social cancer of all. Aside from the economic hardships it creates, it also generates pathology, for it makes people feel useless, which in turn leads to depression, alcoholism and mental illness. Then too, joblessness means more crime and delinquency. (Gans 1977, 42)

In short, unemployment is a devastating personal and social problem. The unwillingness of political leaders to view unemployment as a systemic, rather than an individual problem is a disservice to dedicated workers and misdirects social policies designed to improve the state of the overall economy.

This chapter highlights the significance of unemployment statistics and the ways they are misused, misinterpreted, and politicized in American society. It documents that the original proponents of collecting labor statistics were workers, organized in unions and reform groups, who believed investigations into workplace conditions were an essential step in securing corrective action. They viewed the collection of labor statistics as a way of drawing attention to the problems they confronted, and a channel through which social change might be achieved. But this early reformist intent became heavily politicized, and the activistic aspect of collecting official statistics was eventually abandoned.

With the collection of work-related data resting in the hands of business leaders in some states, and by professional econometricians in others, labor statistics became disconnected from the workers that helped create them. Similarly, the goal of using labor data to secure improvements for workers was abandoned by state bureaus and those who administered them. In some states, the calculated replacement of worker representatives with business-people effectively redirected the purpose of gathering employment-related data. In other cases a change in the philosophy guiding the bureaus was primarily attributable to well intentioned professionals seeking to enhance the "respectability and objectivity of labor statistics." Regardless of the intent, the end result was to essentially transform "labor" statistics into "business" statistics.

Today, labor statistics and the agencies that create them remain estranged from their early association with workers. The unemployment rate now plays a variety of roles, including its use as a leading indicator of the business cycle. The notion that the unemployment rate should be a measure of social welfare, or of its inability to provide jobs for all of those who need them has little currency in contemporary political circles. The way unemployment statistics are calculated and presented is a direct reflection of the way the underlying problem is regarded; individually, not as an outgrowth of the economic system. As this chapter explains, it falls upon workers to "prove" they are unemployed by satisfying criteria that reflect neoliberal definitions of social reality. This chapter maintains that the jobless rate should primarily be a measure of the economic system's ability to provide work for those seeking gainful employment.

In brief, the unemployment rate has been misused and misinterpreted as an accurate gauge of joblessness in the labor force. The following section documents that the creation of the unemployment rate was a product of competing political pressures. Further, in contemporary times, the rate itself has been criticized by actors from competing ideological viewpoints who hope to gain political capital from public assaulting the official reality manufactured by government statistics.

THE EVOLUTION OF UNEMPLOYMENT STATISTICS

Regular publication of unemployment statistics began in 1940. By that point, unemployment had been a recurring problem for more than a century. The unemployment problem was particularly severe during the business contractions that began in 1837, 1873, 1893, 1914, 1921, 1927, and 1929. Before the Great Depression, the classical economics perspective prevailed, emphasizing harmony, equilibrium, and a laissez-faire approach to market regulation. Its fundamental ideological tenets served to minimize concern among politicians and employers about working class unemployment. Only after the most crippling economic downswing occurred did sustained political concern about unemployment emerge. Prior to the 1930s, despite considerable evidence to the contrary, unemployment was considered to be a transient economic phenomenon.

Most professional economists were influenced by mechanistic models that dealt with the relationship between wages and the amount of labor that could be induced into the market by them. As Stewart and Jaffee summarize,

> Economists devoted their attention largely to the refinement of price theory in a general system of equilibrium analysis. It was assumed that the laissez faire doctrine had an inherent harmony, which implied an optimum allocation of economic factors, including labor, through the mechanism of the market; this assumption contributed to an apparent lessening of interest on the part of economists in population and working force matters. (Stewart and Jaffee 1979, 6)

The heavy reliance upon market mechanisms to explain unemployment convinced business leaders and their political allies that joblessness was an entirely temporary and voluntary phenomenon. Workers, they argued, can always obtain jobs if they lower their wage demands sufficiently to be absorbed by the market.

Early Measurement Efforts

The first unemployment data to be systematically compiled reflected the experiences of what were referred to as the "gainfully occupied." This concept was established in the decennial census of 1870 and used until 1930. A gainfully occupied worker was one who pursued an occupation that yielded earnings or their equivalent. Information was recorded for all persons ten years of age and older. No attempts were made to distinguish between the gainfully occupied workers who had jobs, and those who did not. It was simply another method of classifying people, as a way to inventory the general skills of the population.

In the 1880 census, a question to determine the number of workers actually holding jobs was included, but the government never tabulated the results regarding this question. Efforts to determine the residual between the "gainfully occupied and those currently with jobs" was undertaken in 1890 and 1900, but most critics dismissed the results as being unreliable and without any real value. Moses suggests this early concept of the work force was used because it was congruent with employer interests:

> Essentially, the gainfully occupied approach to working force measurement was concerned with assessing the stock of human resources as it related to productive economic activity and its expansion. Hence the focus on employment and occupation—it served employer needs. Although social welfare problems were of concern, they were always cast in the background of the primary orientation, of secondary and minor interests. (Moses 1975, 30)

In the absence of official national unemployment information, some states took the initiative to collect data on their own, as did labor unions and business-backed research groups.

Carroll Wright and the Politicization of the Unemployment Rate

A particularly instructive example of an early attempt to measure unemployment, and how those numbers became politicized, is provided by Carroll Wright and the Bureau of Labor Statistics in Massachusetts. Wright was not allied with the working class. He epitomized the "harmony of interests" philosophy and placed responsibility squarely on workers for their joblessness, seeing no fundamental problem with capitalism as an economic system. Wright's business-oriented approach to measuring unemployment became apparent in his 1878 effort to enumerate joblessness in Massachusetts.

As Bureau chief, Wright instructed town assessors in Massachusetts to include only "able-bodied males over 18" as unemployed and of that group, "those only who really want employment." Wright's position toward workers became clear when he reported the survey results. According to Wright, "the testimony of officials in very many cases was that a large percentage of those out of employment would not work if they could" (Garraty 1978, 108–9). Wright's orientation about who should be counted as unemployed reflected a narrowly-conceived operationalization of a pressing social problem. It was a view that remains alive today as an integral part of the neoliberal economic ideology that dominates social policy. For example, women, whom Wright dismissed entirely in his official counts, are now disproportionately represented among "discouraged" workers. Despite the steadily increasing labor force participation rate among women, there is still a tendency among many employers to view women primarily as mothers and housewives with a less than full-time commitment to work. Wright's criteria for determining who was truly deserving of the label "unemployed" stemmed from what were ultimately arbitrary evaluations. And although the criteria for calculating unemployment has become thoroughly rationalized, they continue to understate the real rate by being magnanimous in identifying who should be counted as employed, and by limiting those who should be categorized as unemployed.

In reviewing Wright's and other scattered efforts to count the unemployed, the American Statistical Association reported in 1909 that "although much has been written and spoken about the extent of unemployment in the United States, no careful attempt has heretofore been made to compile the available facts and furnish a definite answer to the question" (Nearing 1909, 525). In an attempt to fill that void, Nearing culled the available empirical evidence from private and state sources. On the basis of these figures, he concluded that,

> Unemployment is inseparable from the present system of industry: (1) industrial depressions mean extensive unemployment, (2) certain trades are subject to violent fluctuations of demand which result in unemployment . . . industrial uncertainty and personal incapacity make unemployment a constant factor in the life of the average wage-worker. (Nearing 1909, 542)

Another early effort to gauge unemployment was undertaken by the Bureau of Labor Statistics in 1915, with the assistance of the Bureau of Immigration and the Metropolitan Life Insurance Company. Originally limited to New York City, the resources of the Metropolitan Company permitted the canvass to be extended to twenty-eight additional cities. The interview instrument was formulated by the Bureau of Labor Statistics but Metropolitan agents filled in the schedules using their policy holders as survey respondents. Metropolitan's policy holders were numerous, widely distributed, and thus gave the results some degree of credibility when publicly reported.

But labor unions and econometricians sharply criticized the fact that a private insurance company had the most detailed unemployment data. The Commissioner of Labor Statistics wrote, "it is humiliating to note that, in the vitally important matter of unemployment, the facts needed were not available, and that lack of funds rendered the Federal officials helpless to obtain the required information" (Meeker 1930, 387). Commissioner Meeker endeavored to alleviate gaps in 1916 by initiating the regular publication of employment statistics. Although woefully inadequate, the reports became the basis for estimating unemployment for more than two decades.

From the time of the Civil War through World War I, measuring unemployment was a speculative, episodic undertaking. Various estimates were suggested by state labor bureaus, the AFL, the National Industrial Conference

Board. The prevailing belief that unemployment was a sporadic, temporary inconvenience precluded the acquisition of reliable data. As Abramovitz has noted,

> Before Keynes and before the Great Depression, the economists' maps divided all unemployment into three parts; frictional, seasonal, and cyclical, and all of it was "voluntary." Jobless workers, in one way or another, were regarded as holding out for real wages higher than the net revenues which their product would return to an employer. Any single unemployed worker could get some job somewhere by offering to work for lower wages . . . Worker resistance to accepting lower real pay, therefore, was the basic cause of unemployment. (Abramovitz 1976, 24)

In the 1920s, the view that unemployment was voluntary would be vigorously challenged, eventually evaporating entirely amidst the most severe economic contraction in history.

THE 1920S: A DECADE OF MEASUREMENT DISPUTES

Unemployment in 1920–1921 was more severe than the level reached during the previous business decline of 1914–1915. Labor market hardship began to take on large-scale proportions in the fall of 1920. At that juncture, the principal source of official data on unemployment was the BLS' payroll statistics. The Department of Labor augmented the process 1920 when the U.S. Employment Service (USES) would begin gathering unemployment statistics. Each month, the USES published an industrial information bulletin summarizing the employment situation in different industries throughout the country.

Because they tended to rely heavily on employers and other business sources, the USES acquired a reputation for producing reports that overstated positive developments. The first USES Industrial Employment Survey Bulletin, released in January 1921, stated that the number of employed persons in "mechanical" industries had declined by 3.5 million, from 13 to 9.5 million, over the previous year. As dramatic as the 40 percent decline appeared, the figures understated unemployment as it did not include any record of those unemployed in January 1920. The data include joblessness among railroad employees, miners, or others in firms with less than five hundred employees. At the same time, the AFL, in a canvass of 917 cities, reported that 4,000,000 workers were jobless, while the AP reported unemployment at between three and five million (American Labor Legislation Review 1921, 195).

The uncertain nature of counting the unemployed during a serious economic contraction became a source of political embarrassment for the Harding administration. Rather than resolving to produce reliable data, the USES, and others in the administration attempted to minimize the gravity of the unemployment problem. In particular, Labor Secretary James Davis inspired constant criticism from workers due to his department's release of overly-optimistic pronouncements and thinly-veiled attempts to manipulate unemployment information. In the fall of 1922, with a business recovery barely under way, Davis announced that three to four million workers had already returned to their jobs. Davis further contended that three million idle and partially employed workers was the "normal condition in America" (The New Republic 1922, 163). Contained within Davis' assertion was the commonplace social policy implication that idle workers should anticipate little government intervention since their condition is "normal."

When unemployment rose rapidly in 1927, Secretary Davis was again entangled in controversy when he tried to minimize public perceptions about wide-scale layoffs. As joblessness increased, New York Senator Robert Wagner persuaded the Senate to adopt a favorable resolution aimed at collecting unemployment data. The Resolution was a concise statement of the issue:

> Whereas, it is essential . . . to the ultimate solution of the unemployment problem that accurate and all inclusive statistics of employment and unemployment be had at frequent intervals; and whereas . . . no nationwide statistics of unemployment are anywhere available; Resolved that the Secretary of Labor is hereby directed to investigate and compute the extent of unemployment and part time employment in the U.S. and make report thereon. (American Labor Legislation Review 1928, 149)

Complying with the resolution three weeks later, Secretary Davis reported to the Senate that "by the most careful computation methods available, the actual number now out of work is 1,874,050." At greater length, the Commissioner of Labor Statistics, Ethelbert Stewart, wrote:

> The best estimate that can be made from all sources of information available at this time is that the shrinkage in the volume of wage earners, including manufacturing, transportation, mining, agriculture, trade, clerical, and domestic groups, figuring on a base of these employed in 1925, is revealed to be 7.43% . . . 1,874,050 persons. (Bureau of Labor Statistics 1929, 164)

Commissioner Stewart's report explained that 1925 was taken as a base for measuring unemployment because it was an "average" year with respect to labor market fluctuations. He explained that, "in making 1925 the base or 100, it is understood that whatever there may have been of unemployment in the year is ignored, and it is assumed that those who were let out of industry between 1923 and 1924 had by 1925 readjusted themselves." He further suggested that, "it may be said that 1925 was a year in which there was no noticeable unemployment question" (1929, 164).

Stewart's estimates were based on existing data for manufacturing wage earners and railroad employees in 1925. Weeks after Labor Secretary Davis' report, Senator Wagner reported to the Senate that: "One may scan every line and every paragraph of Mr. Stewart's report and nowhere will he find the assertion that there are now only 1,874,050 men unemployed. That statement is purely the product of Mr. Davis' imagination" (American Labor Legislation Review 1928, 151).

And indeed, Secretary Davis' figures were widely challenged. Among others "The New Republic," the American Federation of Labor, and "The Index" published critiques of the data. Eventually, Secretary of Labor Davis conceded that the number of unemployed had exceeded 5.7 million by August 1921, a figure that translated into an unemployment rate of 22.7 percent.

The highly politicized environment surrounding Davis and the United States Employment Service was highlighted again after the onset of the Great Depression. The USES was a primary source of information on the employment picture, but its the contents were essentially short-term predictions about the labor market made by leading employers in major industries. Although they were charged with documenting the entire employment situation, their studies were distorted by neglecting layoffs among recently employed workers. The divergence between the USES reports and economic reality steadily widened as the Depression deepened.

Based on USES data, Secretary Davis reported to President Hoover in early 1930 that the "steel, iron, and automobile industries and virtually every other major industry has shown increased activity since January 6th." In turn, Hoover told the country that "There has been a distinct increase in employment all over the country in the last ten days" (*New York Times* 1930, 1). To the chagrin of the secretary and the president, New York State Industrial Commissioner Frances Perkins, responded to the news with employment reports that were more critical and empirically-based than those provided by the Labor Department. Citing data from over half of the 1,700 business establishments engaged in various industries in New York, she concluded that there had been a "decided decline in employment in the first 18 days of January."

Referring to the USES report, Perkins charged they were not statistical, adequate, nor properly analyzed (*New York Times* 1930). Perkins' claims were substantiated the following month when the Bureau of Labor Statistics reported a further shrinkage in employment. Of the effort to politicize the unemployment rate, the *New York Times* wrote, "justly or unjustly, (Mr.

Davis') attitude of a month ago is now likely to be regarded by the general public as an effort to make the business situation seem better than it was" (1930a).

As unemployment intensified, the Hoover administration continued to struggle with measurement controversies. Particularly noteworthy was a dispute over how to present unemployment estimates made possible by data derived from the 1930 census. Charles Persons, the census' director, differed with the administration over how to interpret numbers on laid-off workers. Persons believed that given the grave economic climate, workers on layoff should be considered unemployed. The Hoover administration disagreed, preferring to categorize them as "employed persons temporarily out of a job." Hoover and his aides insisted that layoffs be presented in a separate category, apart from the unemployed, or not at all. Persons, convinced that the administration was trying to minimize the figures for political reasons, refused to comply with its efforts and resigned his position in protest (Department of Commerce 1978, 24). Due to Persons' resignation and the ongoing dissension surrounding the real number of unemployed, a special canvass was undertaken in 1931. Eventually, Commissioner of Labor Statistics Stewart, owing to continuing differences with the Hoover administration over the presentation of unemployment data, was forced into early retirement (Goldberg 1980, 24).

These intragovernmental controversies highlight the ways official statistics are misused and politicized. The lack of reliable official unemployment figures has been an issue since the United States' first business contraction. But the Great Depression riveted attention to the problem in a way that had not been done previously. And importantly, it provided the necessary context for creating a regular, official statistical series on unemployment. With President Roosevelt's inauguration in 1933, a distinct shift regarding the appropriate degree of government intervention in the economy and the lives of everyday Americans took place. The idea that unemployment was a temporary aberration withered in the context of the most severe economic slump on record. Government policymakers began searching for solutions, for ways to correct the market mechanism's imbalances.

But even the most progressive of Roosevelt's policies were remedial ones, keeping industrial relations intact while attempting to deal with serious and sustained labor market hardship. In turn, this liberal philosophy toward handling the unemployment problem had a profound impact on its measurement.

THE GREAT DEPRESSION: JOB SCARCITY
AND THE LABOR FORCE CONCEPT

During the Great Depression, the concept of the "labor force" was officially operationalized. It included all persons sixteen to sixty-five years of age who were working, or who were unemployed and made an active search to find a job. Labor statisticians also created the "Not-in-the-Labor Force" (NLF) category that accounted for the activities of unemployed non jobseekers such as full-time students and housekeepers. The adoption of this approach to measuring unemployment was influenced by several key factors. First, there was an ongoing political conflict over whom should be counted, manifested in the contentious estimates disseminated by the AFL and the National Industrial Conference Board. Second, bureaucratic imperatives emphasized the need to standardize the measuring process. The introduction of the labor force concept seemed to aid that objective. Finally, there was the influence of pragmatists in the Roosevelt administration who needed to know the minimum number of jobs to create to relieve hardship.

In 1935, Harry Hopkins, the head of the Federal Emergency Relief Administration (FERA), and an advocate of government made jobs over direct relief, became administrator of the Works Progress Administration (WPA). The WPA had been charged with providing jobs for all employable persons on relief rolls. Hopkins believed only those who "really" needed a job should be counted as unemployed. His interest was in knowing how many jobs would need to be created to assist those who really needed work, a demographic group consisting mainly of males with dependents. In the context of extreme job scarcity, the most realistic goal for the administration was to furnish a minimum number of jobs to make it appear progress was being made. Providing jobs to all of the unemployed was not seriously entertained. Government economist Gertrude Bancroft noted,

> There was a constant demand for a measure of the number of jobs that were required in the economy to take care of the jobless and the proportion of those jobless who were actually in need of relief . . . National and local policy at that time required a measure of unemployment that would be equated with the minimum number of jobs needed. There was no demand for a measure of the total labor supply . . . What was important was to distinguish the active, current, "legitimate" jobseekers from all other persons who, under different circumstances, might become jobseekers or might have been jobseekers. (Bancroft 1979, 45–46)

In short, the "labor force" concept emerged from the need among politicians and economists to make a distinction between those who "really needed jobs" from the larger pool of the unemployed. Rather than serving the interests of the jobless, the prevailing ideology ensured that unemployment surveys were designed to satiate employers and government agencies undertaking the count. Worker-oriented concepts were routinely criticized by business leaders and government economists. John Webb, a director of research at the WPA conveyed the conventional wisdom when he stated:

> Seeking work is an activity that can be reported in terms of what the individual is doing at the time of inquiry. It does not depend upon a judgment by the respondent; and it can readily be phrased in neutral terms . . . unemployment measured by means of the seeking work concept distinguishes the unabsorbed portion of the active labor supply in the sense used by the economist in analyzing the supply of, and demand for, workers . . . And finally, its use would provide a basis for standardization of reports. (Webb 1939, 54)

The Bureau's approach of ignoring the "respondent's judgment" may streamline data collection, but it artificially suppresses the number of those officially categorized as unemployed. Some academic observers were supportive of the view that workers should not be trusted to express their own lived experiences in arriving at an official jobless count. Stanley Lebergott captured this perspective when he wrote "wanting work is not a reliably enumerable concept. To ask whether someone 'wants' a job is like asking whether he wants to attend church regularly or pay his taxes in full" (Lebergott 1942, 27).

While professional economists like Lebergott believed workers could not be relied upon to accurately report their status, they viewed employer definitions as objective and value-free. Myers and Webb, two architects of the first regular unemployment series characterized the "bias" involved in using the worker's perspective this way:

> In many cases the loss of the job that initiated the period of unemployment . . . represented the act of discarding of a worker because he was no longer productively employable according to the standards of that particular industry. . . This fact indicates that the employability of the unemployed cannot be determined with any confidence from the worker's own opinion. . . In so far as ability and expressed willingness to work actually report employability, it does so from the point of view of the worker rather than that of the employer. An unskilled worker of 50 years may be without a physical flaw, sound of mind, and actively looking for work, and still be unemployable in the judgment of the employment office that released him. (Myers and Webb 1937, 531)

While casting doubt on the wisdom of using workers' input, the authors' affirm the notion that employers' opinions should be incorporated in the determination of who should be counted. Myers and Webb argue that employers should be the final arbiters of who is able and willing to work, while offering no empirical support to substantiate their belief that these criteria are more "objective."

In short, when the WPA set about to create official unemployment statistics, efficiency and finding a reasonable number of jobs that could be generated by government programs were prioritized over finding the true extent of joblessness. The first report emerged from a contentious debate surrounding who should qualify as part of the labor force. By the time the WPA released their first report on unemployment in January 1940 significant declines were occurring as mobilization was underway for World War II. In many regions of the country, worker shortages became more of an issue than job scarcity.

Because of the marked reduction in unemployment, the monthly canvass was nearly discontinued when the WPA was dismantled. But the Census Bureau salvaged the report and it continues in an expanded format today as the Current Population Survey (CPS). The CPS is the source of raw data for the monthly unemployment rate. Since the first report, the fundamental conceptual framework involved in measuring the unemployed has remained unaltered. There has, however, been a gradual tightening of criteria, resulting in a growing number of the unemployed being classified as "not-in-the-labor force."

THE CONTINUING DEBATE OVER MEASURING UNEMPLOYMENT

The debate over the unemployment rate as a measure of labor market hardship has continued with little respite. On one hand, some have consistently maintained that the rate is a poor indicator of labor market conditions. Moreover, critics note the unemployment rate discloses little about underemployment in the labor force, or about the quality of work. On the other hand, many political conservatives have berated the measure for its alleged propensity to exaggerate economic difficulties. These critics contend that only primary wage earners should be counted when estimating unemployment. The following succinctly expresses this sentiment:

> The housewife who says she is looking for work is a stock problem. The test with her—as with others—is whether she actually did anything about getting a job . . . A teenage girl lolling around the house may have no intention of taking a job as checker at a neighborhood market, though she might be open to an offer from Hollywood. (Business Week 1955, 150)

According to the prevailing conservative narrative, the unemployed are without work because they are too selective, lazy, or so adequately compensated while jobless that they are "not experiencing hard times." Many impassioned arguments from the 1960s, such as the idea that joblessness among young people "aggravate unemployment estimates" are still being heard today (Raskin 1979, 386).

The Gordon Committee

The relatively high unemployment rate in the 1950s generated a sense of disbelief in the statistics among advocates of the free-enterprise system. Much like today, the idea among business leaders was that the unemployment problem was exaggerated because of the participation of particular groups within the labor market. Additionally, charges that workers were lazy or too selective frequently appeared in the media. Chodorov, writing in the libertarian journal, *The Freeman*, expressed the complaints of many on the topic of measuring unemployment:

> The confirmed malingerer . . . will report himself looking for a job while in fact he is thoroughly enjoying his vacation. Another will insist that he is looking for work even though he regularly turns down opportunities . . . he can wait until the right thing comes along. Evidently, the unemployed can afford to be choosey. (Chodorov 1959, 5–6)

In addition to assuming the unemployed were "over-choosey," Chodorov also assailed the modest social programs created to alleviate the hardship that accompanies joblessness. He continued "handouts aggravate the unemployment problem by removing the sting from unemployment" (1959, 7). Along with direct attacks on the statistics integrity, concerns were also expressed about how the US' high unemployment figures provided cannon fodder for Soviet and Chinese propagandists.

Conservatives appeared particularly concerned that the US unemployment rate was higher than most other developed nations. *U.S. News and World Report* opined that the U.S. unemployment rate would look better to the world if it had the same measurement "rules" as Great Britain. In the UK, only those over eighteen who register at a public employment office as unemployed and seeking work are recorded as being out of work. The publication interviewed BLS Commissioner Ewan Clague who affirmed the suspicions of those who viewed the rate as inflated. The interview concluded this way:

> Q. Actually, then, U.S. unemployment is not as bad as the world has the impression it is.

A. No, it isn't . . . our overall figure of, say, 7 percent unemployment includes as unemployed a great many people who are certainly of less weight, in the sense of their importance to their family and to the economy. (Clague 1961, 82–83)

As the 1950s ended, official unemployment statistics were coming under fire from across the political spectrum. One critique featured the long-standing bias against marginalized groups, including women, minorities and teenagers. The official numbers were attacked for including women because many had difficulty viewing them outside of their role as housekeepers. The inclusion of teenagers was criticized because most teens live at home. Counting minorities was critiqued because conventional stereotypes held that minorities were too lazy to actively search for employment. Concerns were also expressed by political and economic leaders that officially recorded high unemployment rates would lead to "unproductive" social welfare programs. Finally, because of the high official rates, many openly expressed concern about preserving the legitimacy of US capitalism before a global audience.

In what would become a key development in the unfolding controversy surrounding official unemployment data, Daniel criticized the rate for being inflated and disparaged the Bureau's personnel. The author maintained that

Through the years the definitions and methods used by BLS to obtain its fig-ures have steadily been altered to magnify the unemployment problem . . . a review of the last 20 years of the BLS' curious operations makes it clear that the claimed rise in unemployment from recession to recession has, to a large extent, been engineered . . . the process of juggling statistics to show progressive dete-rioration in the U.S. economic system. (Daniel 1961, 67–69)

Daniel was publicly criticizing the Bureau, but his real concern was the prospect of a big spending Congress while Democrats were in power. As he expressed it, "the worse unemployment grows—or can be made to appear—the easier it is to push Uncle Sam into new federal spending programs and new controls over the economy" (1961, 69).

Daniel's contentions captured a great deal of publicity. And neither the BLS nor the Kennedy administration took the allegations lightly. On the advice of Labor Secretary Goldberg, President Kennedy appointed a panel of "expert" economists to make a full scale investigation into unemployment statistics, and the BLS' credibility. Meanwhile, the Joint Economic Committee (JEC) appointed a subcommittee to hold hearings on official unemployment statis-tics. The hearings themselves yielded no debate as Daniel declined to appear before the subcommittee. Of lasting significance, however, was the establish-ment of the expert panel—known as the "Gordon committee." In many ways the Gordon committee sharpened the restrictive definitions that the Works Progress Administration had developed.

In testimony before the Joint Economic Committee in1963, the chairman presented a summary of the events leading to the appointment of his committee. Afterward, he proceeded to reaffirm the traditional concepts and definitions used in measuring unemployment. Gordon told the subcommittee,

> The concept of unemployment now in official use is a reasonable one . . . The aim has been to limit the concept to those not working who are actively seeking a job. The Committee approved of this approach. It believed, however, that the present concepts can be sharpened, in particular it urged that an attempt to develop questions that would determine more objectively than is now the case whether a person has taken concrete steps to look for a job. We urged that reliance on subjective attitudes and volunteered information be minimized. We also recommended that the unemployment concept be modified by the placing of a specified time limit . . . in which the jobless person should have taken definite steps to look for work. (Gordon 1963, 5)

Gordon's comments accent the obsession with objectivity that has led to the misuse and misinterpretation of unemployment statistics. Far from being objective, these figures are produced in a highly politicized environment in which the concepts and definitions used in measuring unemployment have come to largely reflect the employers' orientations. Further, chairman Gordon's emphasis on limiting self-reported information highlights the manner in which the worker's perspective was separated from the problem they were experiencing. Definitional straight-jackets eased data collection efforts while obscuring the ability of everyday consumers of the data to fully comprehend the extent of unemployment in the country. The politicized boundaries imposed by the "labor force" concept were re-enforced by the committee and served to minimize attention about a pressing economic problem.

The committee's recommendations had a palpable impact on the measurement of unemployment. Beginning in 1967, the labor force concept was further restricted by the exclusion of fourteen- and fifteen-year-olds, and the inclusion of a rigorous job search requirement. Further, workers out of a job due to strikes were now classified as employed, rather than unemployed. Each of these criteria caused the "official" rate to appear lower than it would have been otherwise. The use of increasingly restrictive criteria continues to manifest itself in the Bureau's operationalization of employment.

THE MEASUREMENT OF EMPLOYMENT
AND UNEMPLOYMENT TODAY

The source of the official unemployment rate is the Current Population Survey (CPS), which began in 1940 as a monthly undertaking of the WPA. In 1942, the US Census Bureau was given the responsibility for administering the CPS and since that time, has periodically expanded the survey. Today, approximately 60,000 households are selected for the survey which translates into roughly 110,000 individuals who are interviewed each month. The sample is designed to represent the entire United States, reflecting urban, suburban, and rural areas. Each month, the Bureau asks each household about the labor force activities of its members. These interviews are conducted either in person or via telephone. During the household's first interview, the census prepares a roster of the household members, documenting demographic characteristics.

The BLS states that because these interviews are the basic source of data for total unemployment, "survey respondents are never asked specifically if they are unemployed, nor are they given an opportunity to decide their status. Instead, their status will be determined based on how they respond to a set of questions about their recent activities." As suggested elsewhere, the BLS' labor force force concept is neither intuitive not consistent with earlier official approaches to calculating unemployment. A critical implication of selecting the labor force concept over the labor supply concept is that the focus is on the individual and their behavior, instead of the labor market. The "labor supply" includes those officially recorded as employed and unemployed millions who are jobless and say they want work but have not made a formal search effort.

Most who do not make an effort believe they are being realistic about the difficulty of finding work during a recession or in an overcrowded job market. Examples include high school graduates who attend college primarily because job opportunities are scarce as well as many "retired" workers. Instead of categorizing them as unemployed, the labor force concept relegates millions of would-be workers to categories such as "in school" and "keeping house." In short, the definitions and concepts used in measuring unemployment places responsibility on the worker, rather than the system.

Counting the Employed

The BLS tallies individuals as employed if they performed any work for pay during the survey week. This includes part-time and temporary workers, as well as those on full-time, year-round schedules. Additionally, individuals are counted as employed if they were absent from their usual job during the survey week because they were ill, on vacation, or for other reasons. These workers are counted as employed because they have a specific position to which they expect to return. The survey excludes the institutionalized population, including those in correctional facilities, mental health care centers, or the armed forces.

Included among those counted as employed are "unpaid family workers," which includes "any person who worked without pay for 15 hours or more per week in a business or farm operated by a family member." Unpaid family workers comprise a small proportion of overall employment, most of whom are either wage and salary workers or self-employed.

Counting the "Unemployed"

The "official" picture of unemployment in the United States is derived from the monthly CPS. Survey respondents are counted as unemployed if they do not have a job, have actively looked for work in the prior four weeks, and are currently available for work. The list of activities that satisfy the Bureau's "looking for work" requirement includes contacting employers or other agencies about work and submitting resumes. The BLS does not count as unemployed those who make "passive" job search efforts such as someone attending a job training course. Workers expecting to be recalled from temporary layoff are counted as unemployed regardless of their job search effort.

The questions used in the CPS are designed to obtain an "objective," rather than a "subjective" account of each individual's labor force activities. Among others, respondents are asked the following key questions: "Last week, did you do any work for (either) pay (or profit)?," "Last week, could you have started a job if one had been offered?," "Have you been doing anything to find work during the last 4 weeks?," "What are all of the things you have done to find work during the last 4 weeks"?

Beyond job-losers, official unemployment figures also include people who quit their jobs to look for other employment, and workers whose temporary jobs ended. First-time entrants in the labor force, and experienced employees looking for jobs after an absence from the labor force round out the ranks of those officially recorded as unemployed.

Between Employment and Unemployment: Workers NILF

The labor force is made up of the employed and the unemployed. The remainder—those without a job who have not made an active search to find one—are recorded as being "not in the labor force" (NILF). Full-time students, those with family care responsibilities, and the retired make up the greatest part of this group. Each month the CPS asks respondents a series of questions to determine the "marginally attached" in the labor force. They include: "Do you currently want a job, either full or part time?," "What is the main reason you were not looking for work during the last 4 weeks?," "Did you look for work at any time during the last 12 months?," and "Last week, could you have started a job if one had been offered?"

To be counted as "marginally attached," respondents must want a job, have looked for work in the past twelve months and be available for work. "Discouraged workers" are an important part of the marginally attached labor force. These jobless individuals believe no job is available to them in their field, or their geographic area. Additionally, workers are recorded as discouraged after they have ceased job search activities for a period, or because they believe they have inadequate human capital to obtain a job. Finally, workers are considered discouraged if they believe employers view them as too young or aged, or who feel they will face some type of discrimination that will prevent them from getting a job.

Since its inception, critics have argued the official unemployment rate is too conceptually restrictive, failing to adequately capture the breadth of labor market conditions. In response, the BLS has developed a set of alternative labor market indices. These measures are published monthly along with the official unemployment rate. Critically, these alternative metrics seldom receive the same media attention as the "official" rate.

ALTERNATIVE MEASURES OF UNEMPLOYMENT AND LABOR UNDERUTILIZATION

The official unemployment rate excludes certain jobless groups, such as marginally attached and discouraged workers. Marginally attached workers are neither working nor looking for work, but indicate they want to work, are available for a job, and have sought employment in the recent past. Discouraged workers, a subset of the marginally attached, provide job market reasons for not seeking employment. Persons working part time for economic reasons (involuntary part time workers) are those who want jobs and are available for full-time work, but have to settle for abbreviated work schedules.

Since the 1990s, the BLS has published six alternative measures of labor underutilization known as U-1 through U-6. The most restrictive measure, U-1, consists of individuals unemployed fifteen weeks or longer. The U-2 measure is comprised of those who have lost their jobs and persons who completed temporary jobs. The BLS' U-3 is the official unemployment rate, expressed as a percentage of the civilian labor force. The BLS' U-4 measure includes the unemployed plus discouraged workers. The U-5 measure includes the unemployed, plus discouraged workers, as well as other marginally attached workers. Finally, U-6 is the most comprehensive measure, comprising all marginally attached workers, plus those employed part time for economic reasons. Using these alternative metrics for December 2020 rendered labor underutilization rates ranging from 3.4 percent of the civilian labor force for U-1, to 11.7 percent for U-6.

Other labor market statistics are critical in fully comprehending the dynamic character of the labor market. Many point to the labor force participation rate (LFPR) as a crucial barometer. This measure reflects the number of people in the labor force as a percentage of the civilian noninstitutional population sixteen years old and over. In other words, it reflects the population that is either working or actively seeking work. The steady decline in the overall labor force participation rate lends weight to the argument that the official numbers are understated. In December 2000, the overall labor force participation rate was 67 percent. By December 2020, that figure had declined to 61.5 percent.

Another key indicator of labor market conditions is the employment-population (E-P) ratio. This figure shows the percentage of the total population that is currently working. Like the labor force participation rate, the employment-population ratio has been declining for years, helping to substantiate the belief that the official rate understates joblessness. Between 1960 and 2000, the employment-population ratio recovered from each recession and continued an upward trend. However, after peaking at 64.7 percent in April 2000, this pattern stopped. In December of 2000, the employment—population ratio in the United States was 64.4 percent. With the onset of the pandemic-induced recession, the BLS reported the employment-population rate reached a modern-day low in April of 2020 when it fell to 51.3 percent. The ratio also experienced the sharpest month-over-month decline in the history of the series (BLS 2020).

When alternative measures are considered, the monthly unemployment rate appears as a dubious measure of labor market hardship. Rather than reflecting empirical reality, the rate is a statistical artifact created by a government agency attempting to present an 'objective' measure of unemployment. Independent sources have helped fill the void left by official measures. Walter "John" Williams, creator of the Shadow Government Statistics website, estimated the real unemployment rate at 26.3 percent, in November 2020 when the BLS's official rate stood at 6.7 percent. Economists, politicians, and the mass media routinely criticized the rate during 2020 often suggesting the "official" numbers were undercounting the unemployed due to the pandemic (Maurer 2020; Kurtz et al. 2020; Brusuelas 2020; Salmon 2020; Iacurci 2020; Furman and Powell III 2020; Wolff-Mann 2020; Lambert 2020).

CONCLUSION

This chapter on the underestimation of unemployment accents several interrelated themes. One is that meaningfully interpreting official statistics requires understanding the concepts and definitions involved in their construction. As Clarence Long noted years ago,

> No statistical magnitude, when finally approximated, is more vigorously challenged than an estimate of unemployment. Yet the challenge is usually made on charges of statistical inaccuracy. It is not often realized that conceptual limits of unemployment are not definite boundaries, but rather are battlefields over which economic and social philosophies are fighting. (cited in Moses 1975, 27)

In the same spirit, the emphasis in this monograph is not on statistical techniques, but on the more fundamental issue surrounding the concepts and definitions that determine how unemployment will be measured. Another important theme is that official unemployment statistics are a product of the social and political environment in which they are the routinely misused, misinterpreted, and politicized. In order to understand the mis-measurement of unemployment, the impact of specific periods on the creation of official statistics, such as the Great Depression, must be considered. The official unemployment rate reflects politicized concepts and definitions and the views of those who embrace the dominant neoliberal ideology in the United States. Specifically, the creation of the "labor force" concept has led to an understatement of the true dimensions of unemployment, even in prosperous times.

This chapter documents that "official" unemployment statistics matter for several pivotal reasons. Official statistics matter because they trigger the government to adopt fiscal and monetary policies to address unemployment. When the official rate underestimates joblessness, societal resources that are mobilized to ameliorate the issue may be insufficient. If the official rate overstates the magnitude of unemployment, it will similarly misallocate resources. Relying on official unemployment statistics, rather than broader labor force measures, risks diverting economic and human resources away from areas with the greatest need.

Second, official unemployment statistics matter because they are misused. Official unemployment statistics are presented as an accurate measure of unemployment and of labor market conditions generally. In reality, the official unemployment rate is a social construction created by neoliberal economic definitions. The official unemployment rate does not count workers without jobs as unemployed unless they make an active search to find work. At the same time, the BLS counts as employed anyone who spends any amount of time performing work for pay.

The misrepresentation surrounding official unemployment statistics matters because they convey a distorted image about the aspirations and behavior of workers. The official unemployment rate underestimates the number of people without work, individuals who tell enumerators they want work "now" and would take a job if one were available. This matters because many ordinary Americans believe the numbers truly reflect the economy, and make important decisions about their lives based upon them. For example, the official unemployment rate sends signals to workers about whether to move for a job in another state.

Further, official statistics on unemployment matter because the data are so thoroughly politicized that partisan groups hold out alternative, competing versions about the extent of the problem. A growing segment of the population are not persuaded by scientific reality and are losing confidence in traditional institutions. Consequently, baseless ideological beliefs are replacing the way ordinary citizens perceive the world. Increasingly, the unemployment rate is seen as something subject to varied, equally valid interpretations. For advocates of the neoliberal laissez faire economy, the conventional way of measuring unemployment overstates the extent of the problem. In contrast, those who are more critical about the role of capitalism are inclined to view the official numbers as understating involuntary idleness.

The fundamental economic and social importance of work creates a sense of urgency about any level of unemployment. As a measure of the overall health of the labor market, the official monthly unemployment rate has consistently failed. The rate has become increasingly divorced from the actual labor market conditions faced by workers. The unemployment rate obscures the blocked aspirations of millions of people who say they want work now. This dismissal of worker interests benefits economic and political elites by undercounting the growing underlying rate of unemployment in the country. The decline in the labor force participation rate of men over time and the increased presence of women and part time workers has produced a large and growing group of unemployed workers from whom employers can draw upon during economic expansions. During downturns, the ever-present prospect that one will become jobless enhances the ability of managers and employers to exact worker concessions.

Despite the unemployment rate's inherent understatement of the true level of joblessness, continuing criticism has been made by those who believe that only some groups in society should be "counted" when calculating the official rate. Historically, the only time unemployment becomes a national concern is when it affects the heads of white, middle-class households. Concepts like "tolerable rates of unemployment," the "full employment unemployment rate," the "normal" and "natural" unemployment rate, as used by economic and political elites, further undermines the idea that full employment is desirable or obtainable.

In describing how the unemployment rate has evolved, several historical influences have been emphasized. The labor force, and thus the unemployment rate's conceptual underpinnings reflect a debate about who should be counted as part of the work force and who is valued in society. As Sar Levitan noted, the issues may appear technical, but they all have important political implications; "there are always winners and losers" (Singer 1979, 9).

The present way unemployment is operationalized makes those who own and control major economic organizations the big winners. The official unemployment rate merely scratches the surface in presenting a comprehensive picture of labor market conditions. It does not reflect the total number of people who are officially unemployed at some time during the year and fails to reflect anything meaningful about the widespread, and growing problem of underemployment. Finally, the rate neglects the millions of people who want jobs but are not counted because they work in the gray zone outside the officially designated labor force.

Chapter Two

The Centers for Disease Control and the Overestimation of Life Expectancy

This chapter traces the historical evolution of the Centers for Disease Control (CDC) from a post-WWII malarial control unit operating on a $10 million budget, to an international organization spanning more than fifty countries and employing more than 20,000 workers.[1] Following an overview of the social history of the agency, the chapter segues to a critical examination of the statistical series for which the agency is best known—average life expectancy.

CDC'S EARLY HISTORY

The origins of the Communicable Disease Center (CDC) are tied to US efforts to control malaria in the southeast part of the United States. The disease was particularly severe in the South and concerns were high that infections would interfere with the military's efforts to mobilize during World War II. The Center was founded on July 1, 1946 as a successor to the Office of Malaria Control in War Areas (MCWA), which had been established in 1942. The organization's founder, Dr. Jospeh Moutin, a career public health official, played a central role in the malaria eradication effort. Moutin advocated that the new agency should focus on all communicable diseases, not just malaria. His high-profile associations with Coca Cola and Emory University, which began when he worked for the MCWA, was an important reason the CDC remained in Atlanta.

The new public health agency, which once occupied a single floor in an office building, paid a token $10 to Emory University in 1947 for fifteen acres on which the CDC's national headquarters were constructed. In 1949,

Dr. Alexander Langmuir joined the CDC to lead the epidemiology branch, a position he held for twenty years. Langmuir, a sometimes controversial figure, saw the CDC as an agency with the potential to do much to advance the cause of public health. Within months of assuming the position as the CDC's chief epidemiologist, he launched the first-ever "disease surveillance" program. The project confirmed his suspicion that malaria, on which the CDC spent the largest portion of its budget, had been eradicated in the United States. Because of the success of Langmuir's methods, disease surveillance became an important cornerstone of the CDC, and it altered the practice of public health.

Partly because of its success in eradicating malaria, the continued operation of the CDC was not a certainty. However, beginning in the early 1950s, several developments unfolded that ensured the organization's survival. The first was the Korean War and the impact it had on the CDC's priorities. In 1951, Langmuir created the CDC's Epidemic Intelligence Service (EIS) as an early warning system against biological warfare. This effort fundamentally altered the agency's original mission as a disease eradication unit and heightened the CDC's national prominence. In addition to the outbreak of the Korean War, two public health developments in the mid-1950s served to cement the CDC's credibility and organizational viability. The first was the discontinuation of the Salk vaccine after the appearance of polio in children who had been inoculated against the disease. By following Langmuir's pioneering "shoe-leather" epidemiological methods, the CDC was able to trace the cases to a contaminated vaccine from a California laboratory. Shortly thereafter, the nationwide inoculation program resumed for first and second graders. Two years later, when a massive influenza epidemic occurred, surveillance methods were again employed to trace the source and national guidelines for the influenza vaccine were developed.

After 1957, the CDC continued to expand as a governmental organization largely through the acquisition of other agencies. The venereal disease program became a part of the CDC in 1957 and with the spread of the communicable disease came the first Public Health Advisors. In 1960, the tuberculosis program became a part of the agency, and the Mortality and Morbidity Weekly Report (MMWR) was merged into the CDC the following year. Later in the decade, the Foreign Quarantine Service (FQS), one of the older Public Health Service units, was integrated into the CDC.

Beginning in the 1960s, the CDC's reputation was enhanced globally as it engaged in public health activities in South America, the Pacific Islands, and Central and West Africa. During this period, the organization was widely credited with aiding in the global eradication of smallpox. In the 1970s and 1980s, the CDC began tracking new disease outbreaks, uncovering the cause of Legionnaires Disease and Toxic-Shock syndrome. Acquired immunodeficiency syndrome (AIDS) was first brought to the public's attention by the CDC in its June 5, 1981 issue of the *Mortality and Morbidity Weekly Report*. Since that time, a significant portion of the organization's budget has been devoted to the prevention and treatment of the disease.

Although its accomplishments were widely recognized, the agency's involvement in a series of research projects ensnarled it in controversy. Perhaps the most damning episode involved the CDC's role in the Tuskegee study of the long-term effects of untreated syphilis in black men. The study, which began in 1932 as a project of the Public Health Service was transferred to the CDC in 1957. Formally referred to as the "Tuskegee Study of Untreated Syphilis in the Negro Male," it initially involved 600 black men—399 who had syphilis, and 201 blacks who were disease-free. The study was undertaken without the participant's consent. Researchers with the CDC told the men they were being treated for "bad blood," a colloquial term used to refer to a constellation of ailments, including syphilis, anemia, and fatigue. In the experiment, the unwitting participants did not receive the proper treatment needed to cure their illness. Instead, syphilitic blacks received only medical exams, meals, transportation, and burial insurance in exchange for their participation. The effectiveness of penicillin as a therapy for syphilis had been established in the late 1940s. But the study was only terminated after it came to the public's attention in the early 1970s. Originally projected to last six months, the study continued for four decades.

In July 1972, Jean Heller, a reporter for the Associated Press broke the story about the Tuskegee Study, generating outrage among the public. Heller reported that "during a 40-year federal experiment, a group of syphilis victims was denied proper medical treatment for their disease. Some participants died as a result, but survivors now are getting whatever aid is possible . . ." (Heller 2017). Three months after her story became front page news, the experiment came to an abrupt halt. By that time, seven men involved in the experiment had died of syphilis and more than 150 of heart failure that may have been linked to syphilis. Seventy-four participants were still alive, but government health officials who began the study had already retired. Senator William Proxmire of Wisconsin, a member of the subcommittee that oversees PHS's budgets, called the study "a moral and ethical nightmare" (Time 1972). In 1997, President Bill Clinton apologized to eight of the survivors, stating "You did nothing wrong, but you were grievously wronged," . . . "I apologize and I am sorry that this apology has been so long in coming" (Waxman 2017).

The CDC was also sharply criticized over its effort to vaccinate the US population against the swine flu in the 1970s. This controversy is particularly noteworthy in light of the emergence of a widespread anti-vax movement in recent years, and the "hesitancy" of many residents to get inoculated against the flu or COVID-19. Much of the contemporary reluctance to receive vaccines can be traced to the failed campaign to protect the public against a strain of the swine flu virus. The government-initiated effort has been chronicled in the annals of public health as a major setback in terms of the perceived effectiveness of the flu shot and Americans' perceptions about publicly funded campaigns to provide vaccines.

In the winter of 1976, a novel strain of influenza caused an outbreak of respiratory infections at Fort Dix, an army post in New Jersey. Initially, public health officials believed the virus was closely related to the 1918 flu pandemic that killed over 100 million people. The fact that Fort Dix had been a point of origin during the 1918 flu pandemic heightened anxieties about the flu strain. The coincidences between the 1918 flu and the new outbreak, along with the virus's "sustained person-to-person spread," prompted health officials to mobilize for a pandemic the following winter. In March of 1976, President Gerald Ford announced an ambitious plan to have every American vaccinated against the virus. Although the World Health Organization was concerned about the new strain, they were more cautious in advocating mass inoculation programs. While the global health organization monitored the unfolding situation, the Ford Administration signed emergency legislation authorizing the creation of the "National Swine Flu Immunization Program." The government-initiated effort succeeded in persuading nearly one-quarter of the US population to receive a vaccine, but serious issues surrounded the program. Because scientists had used a "live virus" for the vaccine instead of an inactive form, it heightened the likelihood that susceptible groups of people receiving the vaccine would experience adverse side effects (Kreston 2013). Additionally, eminent American public health professionals began questioning the program's drain on limited health resources in the nation. As the United States continued on its unilateral path regarding inoculations in a year in which President Ford's reelection was at stake, the continuation of the national vaccine program appeared increasingly politically motivated. Eventually, scientists discovered that the flu was unrelated to the 1918 virus. Those who came down with the new variant of the flu ended up experiencing only a mild illness, while the vaccine that was used resulted in over 450 people developing the paralyzing Guillain-Barre syndrome. Meanwhile, outside the United States, the novel virus failed to blossom into the pandemic the Ford administration expected. As Kreston summarizes, the American public can be notably skeptical of forceful government enterprises in public health, whether involving vaccine advocacy or limitations on the size of soft drinks sold in fast food chains or even information campaigns against emerging outbreaks. The events of 1976 "triggered an enduring public backlash against flu vaccination, embarrassed the federal government and cost the director of the U.S. Center for Disease Control his job." It may have even compromised Gerald Ford's presidential re-election as well as the government's response to a new sexually transmitted virus that emerged only a few years later. What happened in 1976 is a cautionary public health tale, the story of a vaccination quagmire that still resonates in the public psyche and in our discussions about vaccines today.

CDC'S CURRENT CONTROVERSIES

As noted in the opening chapter, the CDC has been in the crosshairs of political controversy surrounding the official number of deaths attributable to the spread of COVID-19 virus. In the early days of the pandemic, a group comprised of four former CDC heads said the United States was contending with two adversaries in its battle to contain the coronavirus outbreak. The first was the virus itself, the other was political leaders who were undermining the CDC. According to this group, the public health agency had been politicized to an unprecedented degree under the Trump administration (Lianne 2020). But the current undercount controversy is just one in a series of conflicts in which the CDC and its statistics have become involved over the years.

As a case in point, the current Director of the CDC has been generating controversy for decades in his career as a public health official. Robert Redfield, appointed in 2018 after the resignation of Brenda Fitzgerald, is a Baltimore-based physician, virologist and professor at the University of Maryland School of Medicine. Unlike other Directors of the CDC, Redfield had no experience as the leader of a public health agency at the time of his appointment. In the early 1990s, he confronted charges of scientific misconduct when it was discovered he misrepresented medical data—exaggerating the benefits of an experimental HIV vaccine he was researching. Although a military investigation cleared him of misconduct charges, Belluz (2018) reports the data he published had to be corrected. Former Air Force Lt. Colonel Craig Hendrix, one of the researchers who first reported the Director's alleged improprieties, said Redfield had either been "egregiously sloppy with data or it was fabricated . . . somewhere on that spectrum, both of which were serious and raised questions about his trustworthiness." Hendrix added "faulty data can lead other scientists to repeat the same mistakes and prompt participants to seek out trials for drugs and vaccines that don't work" (Taylor 2018).

Further, Redfield was widely criticized by public health experts during his tenure with the military's HIV Research Program due to the discriminatory practices he embraced and the religious context in which he placed health issues. For example, in the 1980s he advocated for mandatory HIV testing among those enlisting for military service, the results of which would have been forwarded to health authorities without the recruits' consent. Redfield also called for quarantining and segregating HIV-positive personnel in the military, a policy generally opposed by most medical authorities at the time, including the Surgeon General. In a 2018 tweet, global health expert Laurie Garrett labeled Redfield's HIV program "punitive," noting its emphasis in the 1980s on identifying gay men, finding their lovers and drumming them out of the military.

Redfield has also attracted criticism from many in the scientific and medical communities who are concerned about his extreme religious agenda. A devout Catholic, Redfield sits on the board of an evangelical Christian group known as the Children's Aid Fund. According to Gregg Gonsalves, a professor at Yale and HIV/AIDS activist, in following his ideological orientation, Redfield has called for abstinence-only education for HIV prevention and endorsed nonevidence-based approaches to HIV prevention rather than scientific ones.

Redfield was appointed to replace Brenda Fitzgerald, another Trump-appointee who resigned amid controversy. Fitzgerald resigned after *Politico* reported she had invested in tobacco and pharmaceutical companies after assuming the leadership position at the CDC (Neel 2018; Branswell 2018). As the Director of the CDC, she was the head of the nation's most vigorous smoking-cessation programs, even touring CDC facilities after purchasing tobacco company stocks. Alex Azar, Secretary of Health and Human Services (and a former executive with Eli Lilly), announced the Director's resignation, citing "complex financial interests that imposed a broad recusal limiting her ability to complete all her duties as the C.D.C. director." *Politico* revealed Dr. Fitzgerald's investments in tobacco firms, including Japan Tobacco, Reynolds American, Philip Morris International, and Altria Group. Fitzgerald is not the only eminent public health official to leave under a cloud of controversy in recent years. In 2017, the nation's leading health official, Tom Price (head of the Health and Human Services Department, which oversees the CDC), was forced out of his position after it was disclosed he spent more than $1 million in taxpayers' money for travel, some of which was for chartered flights (Gambino 2017).

Another instance of the CDC's propensity for becoming embroiled in controversy stems from research published in 2018 by four scientists in the CDC's Division of Unintentional Injury Prevention. Seth and colleagues wrote in the *American Journal of Public Health* that many overdoses involving illicit fentanyl and other synthetic black market opioids have been erroneously counted as prescription drug deaths. According to the authors, the "availability of illicitly manufactured synthetic opioids (e.g., fentanyl) that traditionally were prescription medications has increased. This has blurred the lines between prescription and illicit opioid-involved deaths." According to the authors, the CDC has traditionally included synthetic opioid deaths in estimates of prescription opioid deaths, but with the increasing accessibility of illicit fentanyl, the official number of deaths attributed to prescription drugs could be significantly inflated.

According to the CDC's official estimate, 32,445 Americans died from pain medication overdoses in 2016. But when overdoses involving illicit fentanyl are taken into account, the number drops to 17,087 deaths. The editorial by the CDC researchers carries a disclaimer that their views "do not necessarily represent the official position" of the CDC. Anson (2018) notes that the misinterpretation and politicization of opioid overdose numbers has been a source of concern among public health officials, who have issued significantly different accounts of the number of Americans who have died from opioid deaths in recent years. In 2016, within a one week period, the CDC and the White House Office of National Drug Control Policy released three different estimates of the number of Americans who had died from prescription opioids the previous year, ranging from a low of 12,700 to a high of 17,536.

According to President Trump's opioid commission, the data tracking system "does not have sufficiently accurate and systematic data from medical examiners around the country to determine overdose deaths both in their cause and the actual number of deaths." As Anson (2018) suggests, finding accurate numbers is necessary in order to find real solutions to the overdose crisis. The current system has led to a health care climate in which doctors are too intimidated to prescribe opioids and patients cannot receive the treatment they legitimately need. The CDC researchers wrote, "inaccurate conclusions not only mask what's driving the overdose crisis—they mask the solutions too. Obtaining an *accurate count* of the true burden and differentiating between prescription and illicit opioid-involved deaths are *essential to implement and evaluate public health and public safety efforts.*"

Oliver et al. (2019) have echoed concerns about the CDC and the reliability of their opioid overdose (OD) statistics. As the authors note,

> The primary statistic of concern is the number of opioid OD deaths, which have undeniably increased over the last two decades. Between 1999 and 2017 . . . more than 218,000 people have died in the U.S. from ODs related to prescription opioids, and OD deaths involving prescription opioids were five times higher in 2017 than 1999. (CDC 2018)

The CDC authors highlight that while these numbers are familiar to many lay and professional observers, few are aware that the number of opioid prescriptions and prescription opioid–related deaths peaked in 2011. Declines were recorded during 2012–2014, when the numbers were recalculated to separate deaths from prescription drugs apart from those attributed to illicit opioids. In short, although opioid-related deaths continue to rise significantly,

the critical factor in the recent rapid rise in fatalities is the use of illicitly manufactured fentanyl and heroin (Lilly 2018). Seth and colleagues emphasize
that this nuanced narrative of opioid use and the cause of overdose deaths is
seldom conveyed in authoritative publications, leading to a misinterpretation
of the role of prescription opioids in understanding the current public health
crisis. Further, they note some data published by the CDC can be misleading
and the agency is less than fully transparent in how they calculate official
overdose statistics. O'Donnell (2017) adds that the CDC does acknowledge
an ongoing inaccuracy in death data collection that may falsely raise prescription opioid mortality rates, leading to a misclassification of some fatalities
due to heroin overdoses.

Despite the dubious nature of the data, many governmental publications,
such as those from the CDC, have used the "official" rise in death rates to
justify calls for reducing the number of opioid prescriptions being written by
medical practitioners. One of the reasons official statistics are socially significant and deserving of critical scrutiny is because they drive social policy. In
this case, overdose statistics have been used to address a public health concern
in a way that ignores individual differences in treatment modalities (Dowell,
Haegerich, and Chou 2016; NIDA 2018; Substance Abuse and Mental Health
Services Administration 2018; U.S. Food and Drug Administration 2018).
The way dubious official statistics can be misused to influence social policy
is further highlighted in the following examination of the CDC and its official
statistics on life expectancy.

THE CDC AND LIFE EXPECTANCY STATISTICS

Official life expectancy statistics are socially and politically important
because of their social policy implications. But it is the CDC's emphasis on
conventional ways of measuring and presenting life expectancy data that
leads to a widespread misinterpretation of the statistics. The sharp increase in
average life expectancy *at birth* over the past century is a statistical artifact
owing to society's ability to keep newborns alive during their first year; by
reducing the infant mortality rate. Little of the increase in average life expectancy is due to heroic, high-technology medicine extending our lifetimes.
Whereas average life expectancy *at birth* has grown by more than thirty years
in the past century, the median number of additional years a sixty-, seventy-,
or eighty-year-old can expect to live has advanced nominally. Age-specific
death rates are less publicized and clearly show that people are not living
as long as the official life expectancy figure suggests. The decision to use
at birth vs. age-specific data reveals a recurring theme in the manuscript;
the government often acknowledges the official statistics they publish are

incomplete, easily misinterpreted, or are easily manipulated for political purposes. As with official unemployment data, the government reveals their measure is inadequate by publishing the U-1 through U-6 data series on labor force utilization. Likewise, chapter 3 on official crime statistics signals the government is aware the crime index is insufficient by publishing both the National Crime Victimization Survey, *and* the official Uniform Crime Report (UCR). Similarly, the CDC publishes an official life expectancy at birth figure as well as a wide range of data that renders a more complete understanding of longevity trends. Finally, in chapter 5 on the Census Bureau, the government acknowledges its failure to enumerate the entire population when it publishes its official estimate of the undercount from the census in the aftermath of the decennial canvass.

This chapter does not impute nefarious motives to the CDC, or its statisticians. Rather, it highlights how life expectancy data—showing an inexorable increase among the American population—has already justified raising the full retirement age from sixty-two to sixty-seven, and in the process, subverted the original intent of the Social Security retirement program—paying out more in benefits to well-to-do recipients—than to low-income retirees. The chapter documents how the same misleading life expectancy data is driving private companies to alter pension plans to increase the retirement age for their beneficiaries. Upon critical inspection, the manner in which life expectancy data can be used to influence personal decisions also becomes transparent. Misleading data about how long one expects to live are misused because they factor into decisions people make every day regarding key aspects of their life, including those involving marriage, divorce and child-rearing. Before specifically discussing the social and fiscal implications surrounding the use of dubious life expectancy data, this chapter operationalizes key terms and identifies pivotal issues regarding life expectancy.

DEFINING LIFE EXPECTANCY: "PERIOD" VS. "COHORT" LIFE EXPECTANCY AT BIRTH

In discussing the myth of life expectancy and its social and political implications, it is important to define key concepts, beginning with "myth" and "life expectancy." A myth is a widely held belief, one that often connotes a false idea. This chapter demonstrates that official life expectancy data lends itself to the creation of such a myth; namely, that our longevity has advanced more than thirty years since 1900. Conventionally, "life expectancy" refers to life expectancy at birth or (LEB), which can be operationalized in two ways: "cohort" life expectancy at birth, and "period" life expectancy at birth. Of the two, "period" life expectancy is most commonly used. Period life tables are constructed by applying the age-specific death rates of a given population for a given year divided by a hypothetical cohort of 100,000 newborns.

MISINTERPRETING LIFE EXPECTANCY STATISTICS: LIFE EXPECTANCY AT BIRTH VS. LIFE SPAN

Several issues contribute to the misuse and misinterpretation of life expectancy statistics. To begin, many people confuse *life expectancy* with *lifespan.* The latter term refers to the average maximum length of time a member of a species can be expected to survive. For humans, the maximum lifespan appears to have peaked in the 1990s at around 115 years. In 2016, Oshansky corroborated the work of many demographers in suggesting that life expectancy increases were likely to stagnate in coming years. Additional support can be found in Dong, Mulholland and Vijg (2016), whose global data show that improvement in survival with age tends to decline after age one hundred. In noting that the age of the world's oldest person has not increased since the 1990s, the authors conclude that the maximum lifespan of humans is fixed and subject to natural constraints.

UNDERESTIMATING LIFE EXPECTANCY IN THE DEVELOPING WORLD AND AMONG ANCIENT ANCESTORS

Much popular and professional literature on life expectancy tends to underestimate the lifespans among people in the developing world and those who lived long before the contemporary period. It is the use of the life expectancy *at birth* concept that tends to distort our thinking about the lifespans of these groups. Research by Kanazawa (2008) shows that when infant mortality rates are controlled for, the average lifespan among hunter-gatherers was between seventy and eighty years, a figure comparable to that in the industrialized west. Holloway argues that "our ancient ancestors were not dropping dead at age thirty-five, and some would have even been blessed with long and healthy lives" (2014). Similarly, Radford asserts the idea that our ancestors routinely died young (say, at age forty) has no basis in scientific fact. Kanazawa adds that . . . most adults, both in our ancestral past and in many developing nations today, live to be about as old as people in western industrialized nations (2008).

Volk (2008) estimates that as many as half the children during our evolutionary history, and as recently as the eighteenth century, may have died before the age of twelve. His research shows that while life expectancy at

birth is much lower in our evolutionary history than in contemporary industrialized nations, life expectancy at fifteen years and at thirty years is not dissimilar "adults everywhere and at all times, live to be about the same age."
Similarly, Laden (2011) notes many people misinterpret and underestimate
the life expectancy of past populations. According to *Condensed Science*, the
lifespan of hunter-gatherers is not as low as commonly thought and in many
respects rivals that of the industrialized world (2011).

THE IMPACT OF REDUCING INFANT MORTALITY ON LIFE EXPECTANCY

Average life expectancy is misunderstood because the impact of reducing
infant mortality on creating "advances" is largely ignored. Life expectancy at
birth is not a useful statistic if the goal is to compare the health and longevity
of adults. To illustrate the problematic nature of using LEB data, consider the
case of a couple with two children; one that dies during year one, while the
other survives until year eighty. It would be accurate to state that the average age of their children was forty, but it would be substantively meaning-
less. What makes more statistical sense is to examine life expectancy at age
five, twelve, or at some later year, say fifty, sixty, seventy, or eighty. A more
nuanced understanding of life expectancy is revealed by examining age-
specific mortality data life.

As age-specific life expectancy data published by the CDC reveal, in 1850,
a sixty-year-old white male could anticipate living another 15.6 years. By
2011, that number had advanced less than six years to 21.5 years. Similarly,
sixty-year-old females in 1850 could expect to live another 17 years, while
their counterparts in 2011 could expect a gain of 24.5 years. If we look at
similar parameters for 1900, we see that the improvement remains at less
than six years for white men compared with 2011, and for white females, the
improvement is less than a decade. Although age-specific death rates clearly
show improvement in life expectancy over time, it is far less than the widely
heralded thirty year average suggested by the CDC.

OVERESTIMATING THE ROLE OF MEDICAL SCIENCE IN INCREASING LIFE EXPECTANCY

Many Americans believe the gain in average life expectancy stems from
medical intervention; a byproduct of high-technology medicine, skilled
practitioners, and pharmaceutical products. Yet, the preponderance of evidence suggests most of the gain in life expectancy took place before the

medical-industrial-complex was established, before major research and teaching medical universities had been created, and before most advanced medical interventions were available. The disease epidemics that wreaked havoc in the Middle Ages were not quelled by mass inoculation campaigns. Instead, enhanced hygiene, the impact of an environment characterized by clean air and water, rather than industrial pollutants, are responsible for increased longevity.

According to McKinlay and McKinlay (1977), medical intervention accounted for less than 5 percent of the increase in the average life span in the United States since 1900. In a similar vein, Taylor (1979) states that "almost 80 percent of the reduction in infant mortality between the 1880s and the 1970s occurred prior to the 1930s." Survival rates for infants steadily advanced long before the use of pharmaceutical drugs or vaccines were widely adopted. Further, Sharpe (1988) notes that "mortality for virtually all the infections was declining before, and in most cases, long before specific therapies became available . . . the impetus to better health from the mid-nineteenth century onwards can therefore be directly traced to public health measures and social legislation that improved the living standards of working people." Moreover, higher wages and welfare benefits made it possible for the poor to eat properly and public health measures radically improved conditions in densely-populated urban areas, particularly with the provision of clean water supplies, sanitation, sewerage and new housing. Susceptibility to infections diminished radically as nutrition, housing, hygiene and general living conditions improved. According to Ramsay and Edmond (1967) it is a "widely held fallacy that mortality from infectious disease only commenced to fall with the advent of modern pharmaceutical agents."

The notion that public health measures are key to insuring longevity can be seen in analyses of Japan's life expectancy rate, most commonly ranked as the highest in the world. According to Murray (2011), the rise of thirty years in expected lifespan from 1947 began with the rapid economic growth of the nation in the late 1950s and 1960s. During this period, the government invested heavily in public health, introducing universal health insurance in 1961, and free treatment for tuberculosis and programs designed to address childhood diseases related to intestinal and respiratory infections. In addition to public health measures, Murray emphasizes factors such as a heightened emphasis on hygiene in all aspects of life, including a low fat diet, and being health conscious in extending average life expectancy among the Japanese.

THE OFFICIAL DECLINE IN LIFE EXPECTANCY

According to data published by the CDC in 2016, life expectancy in the United States fell for the third consecutive year in 2014 for white Americans and was unchanged for all population groups as a whole. At the same time, the agency reported a drop in the infant mortality rate in 2014, to a historic-low of 5.8 deaths per 1,000 births (among children aged one year or less). A white American born in 2014 could expect to live to 78.8 years; a decline from 78.9 years. Non-Hispanic white women in the United States live longer than white men and African Americans of both sexes., with a life expectancy at birth of 81.1 years in 2014; a slight decline from 2013, when their life expectancy was 81.2 years. Reflecting on the 2014 data, the author of the CDC report stated, "basically, we're back to where it was in 2009" (Arias 2016). The decline in life expectancy for white women was the first since the CDC began keeping records nearly one hundred years ago. Arias found that increases in unintentional injuries, suicide, and chronic liver disease were major causes for the decline. In another report, published in the *Journal of the American Medical Association*, Anne Case (2016) referred to white people dying in middle age as "deaths of despair" (a phrase quickly frequently repeated in the mass media). In an interview with National Public Radio, she explained that her research showed that deaths of despair are concentrated among white people with less education—high school dropouts and people who never went to college. . .It's possible working-class whites have lost the narrative of their lives" (cited in Kodjak 2016). The 2016 CDC report did report improvement for some groups—African Americans gained 0.4 year of life expectancy in 2014. The largest percentage gain in life expectancy at age sixty-five was among Hispanic males, whose expected increase in life expectancy advanced by 0.3 years.

According to Arias, reversals, even small ones, are unusual for wealthy nations. In the United States, life expectancy overall has been stagnant since 2012, remaining below that of many European countries, as well as Canada, Australia and Japan. The stagnation follows decades of steady increases from 69.7 years in 1960 to 76.8 years in 2000 and 78.7 years in 2010. Although causes of the recent stagnation cannot be identified with precision, many demographers believe that deaths from illegal drug use, suicides, and obesity are the key factors contributing to the trend. A study by Case and Nobel Prize laureate Angus Deaton (2015), suggests that mortality among middle-aged white Americans has been on the rise for fifteen years because of alcohol, drug abuse and suicides, particularly among disadvantaged members in society.

Worldwide, public officials and academicians have identified a trend toward a shortened lifespan, including Yvette Cooper, a former UK health minister. In a speech in 2000, she explained that there was a risk that the recent gains in life expectancy might be lost due to increased mortality associated with childhood obesity, and diseases associated with obesity (specifically heart disease and Type II diabetes). Similarly in March of 2004, Richard H. Carmona, then Surgeon General of the United States, told a Senate subcommittee that because of increasing rates of obesity, unhealthy eating habits, and physical inactivity, we may see the first generation that will be less healthy and have a shorter life expectancy than their parents (Staropoli 2015).

An article in the *New England Journal of Medicine* (Olshansky 2005) stated that if left unchecked, obesity will have a negative effect on life expectancy. It added that the possibility of an infectious disease pandemic and only modest medical advances in coming decades could lead us toward a decline in life expectancy. In sum, many highly respected voices are outlining a narrative in which the long taken-for-granted increase in life expectancy, especially among those in the United States, is being reversed. This trend—as will be highlighted next—is in large part due to the inordinate degree of social inequality that characterizes the United States.

LIFE EXPECTANCY DISPARITIES: INEQUALITY BY RACE

Since the government began publishing statistics on life expectancy, there has been a substantial gap between the life expectancies of black and white Americans. Although that gap has diminished since 1970, in 2010, white Americans could still expect to live four years longer on average than black Americans. Between 1970 and 2010, overall life expectancy at birth increased from seventy-one to seventy-nine years. For whites, the advance during this period went from seventy-one to seventy-eight; among blacks, the increase went from sixty-four to seventy-five years. The difference in life expectancy can be attributed to higher death rates for the black population from disease, cancer, diabetes, and perinatal conditions. Homicide was an additional factor contributing to shorter life expectancies for black males, whose life expectancy was 4.7 years lower than that of white males. Meanwhile stroke was a more common cause of death among black females, contributing to the 3.3 fewer years of life they can expect to live, compared with white females.

LIFE EXPECTANCY INEQUALITY BY CLASS

According to Chetty (2016), the richest 1 percent of American men live 14.6 years longer, on average, than the poorest 1 percent. For women, the average difference is just over ten years. Chetty's research, which involved an analysis of more than 1.4 billion federal tax returns, revealed that between 2001 and 2014, life expectancy rose by roughly three years for the richest 5 percent of the population, while for the poorest 5 percent, there was no increase. To illustrate the gap, Chetty notes that for the richest Americans, the advances in longevity recorded over the past fifteen years equates to curing cancer (eliminating all cancer deaths would add three years to the average lifespan in the United States). He adds that men in the bottom 1 percent have a life expectancy comparable to the average life expectancy in Pakistan or Sudan. Additionally, the study sheds light on the impact of region or area, finding that the poorest citizens of some cities (New York, L.A., and Miami) lived nearly as long as the richest 1 percent in other areas (Las Vegas or Tulsa). Chetty states the statistics have clear implications for Social Security and Medicare. The fact that people don't live as long means they are paying into the system without getting the same benefits, a fact that needs to be considered in any discussion about raising the retirement age (Zarroli 2016).

According to a Governmental Accountability Office (GAO) report (2016), those who delay retirement—and collecting Social Security—are disproportionately well-off, and by age seventy, begin earning larger checks than those who are less well-off. Over time, a growing share of all Social Security benefits will go to people with higher incomes while a shrinking share will go to those with lower incomes. According to a study by economists at the National Academies of Sciences, Engineering and Medicine (NAS 2016) the richest beneficiaries born in 1960 will accrue benefits worth $522,000 (in 2010 dollars), while the poorest born in 1960 will accrue benefits amounting to less than $400,000.

The National Academies report discusses raising the retirement age for Social Security, an idea advocated by several Republican candidates seeking their party's nomination for the US presidency in 2016. According to the report, raising the full retirement age by three years, from sixty-seven to seventy, would reduce the value of lifetime S.S. benefits for the wealthiest group of fifty-year-olds by 20 percent, but for the poorest group, benefits would decrease by nearly 25 percent. Since the poorer group cannot expect to live as long, forcing them to wait to begin drawing their benefits in full would eliminate a relatively larger portion of their retirement. Committee chairman Peter Orszag, a former senior official in the Obama administration, stated, "It's a common refrain to say, 'life expectancy is increasing, so we should raise the normal retirement age under Social Security. . .the problem with that is, it's true that average life expectancy is increasing but that's disproportionately because high earner's life expectancies are increasing" (Ehrenfreund 2015).

Bosworth et. al (2016) corroborates the idea that growing disparities in life expectancy have dire consequences for the less fortunate in their retirement years. Their study looked at life expectancy at age fifty from 1984 to 2012, and found that for men born in 1920, there was a six-year difference in life expectancy between the top 10 percent of earners and the bottom 10 percent. But for men born in 1950, that figure had more than doubled, to fourteen years. Among women, the gap between the highest and lowest-income individuals grew to thirteen years, from 4.7 years. Overall, life expectancy for the bottom 10 percent of wage earners improved by just 3 percent for men born in 1950 compared with those born in 1920. Among the top 10 percent of wage earners, however, the increase was nearly 28 percent. Elizabeth Bradley, a professor of public health at Yale, remarked of the gap, "at the heart of the disparity are economic and social inequities, and those are things that high tech medicine cannot fix" (Tavernise 2016).

LIFE EXPECTANCY INEQUALITY BY GENDER

According to the CDC, white women born in 1930 had a life expectancy of 61.4 years whereas for white men, the figure was 59.7 years. By 2010, life expectancy at birth for white females reached 81.3 years. Meanwhile, black women born in 1930 could expect to live 49.2 years and black men 47.3 years. By 2010, the black female life expectancy figure had reached 78 years, while the figure for black men had reached 71.8 years.

Across the globe, life expectancy is higher for women than for men, with much regional variation. The differences between the sexes are greater in developed nations than in the developing world. Generally, the divergence in life expectancy between the sexes is greatest at birth and narrows significantly by age sixty-five. For individual women, the importance of the life expectancy disparity often means outliving one's spouse, living alone, and poorer economic conditions as diminished resources are stretched to sustain additional years of life. As Lisa Berkman, noted "we (the US) now ranks at the bottom of the OECD countries. This wasn't true 30 years ago; it wasn't true 50 years ago. What's happened is that every other country has improved substantially and we've improved a tiny bit. So life expectancy in the US has improved over time, but it's improved so much less than all other countries that we're now behind." Berkman further states that "while the best off, most educated, wealthiest Americans have continued to experience increases in life expectancy, those at the bottom have not only stagnated, but in some cases have actually lost years of life expectancy over time" (Huffington Post 2014).

As the U.S. National Academies of Sciences study makes plain, rich women will live more than thirteen years longer and collect tens of thousands of dollars more in federal retirement benefits than poor women. Furthermore, despite gains in life expectancy among higher earning women, poor women are actually losing ground. The stress and social inequities associated with poverty manifest themselves in both mental and physical health disparities. Research shows the mental stress of being poor is a major reason low-income people are more likely to have high blood pressure, elevated cholesterol, and to become obese or diabetic. Additional evidence suggests that countries with higher levels of economic inequality have significantly greater disparities in health outcomes ranging from life expectancy and premature death to infant mortality and obesity; even modest increases in wealth inequality are associated with more than double the cumulative risk of death over a twelve-year period. These inequalities have significant implications for social policy as longer-living wealthy people garner a greater share of scarce resources over the course of their lives. In contrast, poorer people who die early are increasingly paying for benefits they will never receive. Instead, the benefits of poorer individuals are being redistributed to higher income individuals, reinforcing income and wealth inequality. As Public Health Watch notes, the current life expectancy patterns may be of particular importance to young women just starting their careers who will see a growing part of their federal benefits diverted from the lowest to the highest earners (PublicHealthWatch 2015).

Increasingly, evidence affirms that class is more critical than gender when it comes to understanding longevity. For example, for the first time since official records were collected, affluent men in England and Wales were recorded outliving women of average means. In 2015, the Office for National Statistics (ONS) reported that "higher managerial and professional men" could expect to live for 82.5 years, one month longer than the average woman. According to the ONS, the gap between the sexes has narrowed dramatically in the past thirty years. Their data also shows that social class differences are a complicating factor, with males in lower income occupations having a life expectancy roughly six years less than those in higher income jobs. According to the ONS, since 1970, inequalities in life expectancy by income have widened for both sexes. As reported by Collinson (2015), in 1997–2001, the poorest women lived, on average, 3.8 fewer years than the richest women, but that figure has since widened to 5.3 years, the biggest gap national statisticians have ever recorded. In all, data from Public Health England on life expectancy in England documents more than one hundred areas, districts, or neighborhoods in which men are now outliving women, in some cases by as much as 5.5 years (Daily Express 2014).

In sum, although women worldwide continue to outlive men, it is increasingly clear that the role of social class is the dominant factor in explaining disparities in life expectancy. Specifically, women have now begun to lose the historic advantage they held due to widening income and wealth inequality. The social policy implication is plain: over time, poor women will be inordinately burdened paying for benefits received by wealthy recipients.

LIFE EXPECTANCY INEQUALITY BY NATION

According to the World Health Organization (WHO), a girl who was born in 2012 can expect to live to around seventy-three years, and a boy to the age of sixty-eight, an increase of six years over the worldwide average for a child born in 1990. According to the UN, life expectancy has increased, on average, nearly three months per year over the past 160 years. The WHO's Director-General, in a 2014 press release stated that: "An important reason why global life expectancy has improved so much is that fewer children are dying before their fifth birthday. But there remains a major rich-poor divide; people in high-income countries continue to have a much better chance of living longer than people in low-income countries" (WHO 2014).

In 2015, Japan had the highest estimated life expectancy in the world at 84.7 years. For girls, life expectancy at birth reached 86.8 years, and for a boy, 80.5 years. The WHO's report showed that in 2012, a male born in a high-income country can expect to live to roughly sevnty-six years—sixteen years longer than a male born in a low-income country. The disparity is even greater for females; a gap of nineteen years separates life expectancy in high-income countries (82 years) compared with low-income countries (63 years).

CONCLUSION: THE SOCIAL AND PERSONAL IMPLICATIONS OF OFFICIAL LIFE EXPECTANCY STATISTICS

This book's examination of the misuse, misrepresentation, and politicization of official statistics is amply illustrated in this investigation of the CDC's life expectancy data. Official life expectancy statistics are socially significant because of the role they have played, and continue to play, in the debate surrounding efforts to "reform" Social Security and private pensions. To reiterate, age-specific mortality rates reveal that Americans today are living six to eight years longer than they were a century earlier, not thirty years longer as the CDC suggests. The political argument that the retirement age should be raised is based on numbers widely misinterpreted by the public and ones that are misused when adopted as a foundation for creating social policies.

Nonetheless, the full retirement age for Social Security beneficiaries has already been raised from sixty-two years to sixty-seven years based on official increases in life expectancy. Similarly, many private pension plans are undergoing a fundamental change. According to Kadlec, (2014), a typical pension has only 85 percent of the funds it needs based on the most recent mortality rates available. The majority of large companies have frozen or changed their pension plans in order to reduce their financial risk, while shifting workers to 401(k) plans. According to industry observers, a growing number of employers will abolish their traditional pensions altogether and offer workers a lump sum settlement instead.

The decision to raise the full retirement age for social security recipients subverts the original progressive, redistributive intent of the nation's retirement program and other social welfare programs. Because they are living longer, wealthy beneficiaries increasingly account for a greater share of government expenditures for Social Security benefits.

Although the specific reasons for the widening gap in life expectancy by class are complicated to explain, the policy implications are plain. When the gain in expected life expectancy is increasingly concentrated among the wealthier members of society, it is misguided to require them to bear the main burden of an aging society. Lifting the retirement age only makes sense if gains in life expectancy are enjoyed equally by rich and poor alike. It is unwise, both politically and morally, to ask the poor to wait additional years for retirement benefits when a disproportionate share of the improvement in life expectancy is concentrated among the affluent.

Finally, an examination of official life expectancy statistics reveals additional ramifications that ensue from their misuse and misinterpretation. Consider the line of research within evolutionary theory known as "life history" that emphasizes how people's beliefs about "life expectancy" influences major personal decisions. Krupp's (2012) research suggests decisions relating to marriage, divorce, abortion, child rearing and pursuing higher education may all be influenced by how long people believe they will live. According to Krupp, "life expectancy might be driving all of these major decisions. The longer someone expects to live, the more time they will invest in education. If life expectancy is short, someone may decide to get married and have children sooner, or stick with the partner they are currently with rather than seek a divorce." Due to the omnipresence of official life expectancy statistics, the personal implications of Krupp's research are far-reaching, particularly for less affluent individuals and their retirement decisions.

NOTE

1. Portions of this chapter were originally published in R.E. Parker and Vianett G. Achaval, "The Myth of Increasing Life Expectancy and its Social Policy Implications." *Journal of Sociological Research* 8(2): 17–31.

Chapter Three

The FBI and the
Miscalculation of Crime

The FBI's annual Crime in the US (CIUS) report attracts the attention of laypeople, academics, and public officials. And many consumers of the official crime data are critical of the Bureau's efforts. According to Walsh and Jorgensen "all social statistics are suspect to some extent, and crime statistics are perhaps the most suspect of all. They have been been collected from many different sources in many different ways and have passed through many sieves of judgment before being recorded" (2017). In this chapter, the social, economic and political implications of the Bureau's crime data is a central concern. Particular emphasis is paid to the way the FBI's official crime statistics are misinterpreted as presenting the complete picture of crime in the United States. Further focus is devoted to the way official statistics are routinely politicized to advance the wishes of actors who misuse the data to support their ideological position. As will be embellished next, official crime statistics have long been politicized (Clark 2018), particularly among actors with specific social agendas, such as gun advocates (Eads 2018).

Two overarching issues with respect to the creation and circulation of official crime statistics occupy our attention in this chapter. The first derives from the way illegal behavior is dichotomized by the FBI into "Part I" and "Part II" criminal offenses (Barnett-Ryan et al. 2014). When the Bureau releases its annual crime report, they are referring to data contained within Part I of the Uniform Crime Report (UCR). Other types of crime, which may be more harmful to larger groups of people and of greater overall cost to the US economy—such as corporate, white-collar, and computer-enabled crime—are neglected in this accounting. Consequently, those types of crimes seldom receive the kind of incessant media coverage garnered by more sensational, fear-inducing street crimes.

The second key concern regarding official crime statistics is reflected in the differences between the UCR and the National Crime Victimization Survey

(NCVS) (US DOJ 2017). The methodologies used for assembling these two series of statistics are dissimilar and, not surprisingly, often yield disparate results. The Uniform Crime Report relies upon data submitted voluntarily to the FBI by nearly 20,000 agencies at every level of law enforcement across the United States, not all of whom utilize the same data-gathering techniques. In contrast, the NCVS is a national survey of more than 70,000 households conducted biannually that asks respondents how often they have been the victims of criminal behavior. The NCVS is an important resource criminologists use to investigate the "dark figure of crime"; referring to crimes that go unreported or underreported.

In the early stages of the coronavirus outbreak, the FBI's ability to track crime for the purpose of organizing statistics for its "Crime in the US" report was severely compromised. As with the other agencies covered here, their official statistics were widely criticized during the pandemic. In particular, the Bureau was roundly critiqued for undercounting certain categories of crimes. Many accounts from around the nation suggested that an uptick in domestic violence, including child abuse was being underestimated during the pandemic's first year. As Sawyer noted, the pandemic engendered several important implications for those generating official crime statistics, including altered policing practices (2020). During the early days of the outbreak, many law enforcement officers were directed to limit unnecessary contact while on duty. This led to a decrease in crime rates stemming from fewer routine face-to-face and street interventions. A further complicating factor was the reluctance of individuals involved in domestic violence scenarios to report incidents during a viral outbreak. The desire of individuals to avoid arrest does away with any record of the behavior and artificially suppresses official crime rates.

Finally, the pandemic posed a challenge in accurately recording crimes as an epidemic of online offenses were being committed during the early stages of the public health crisis. The deputy director of the FBI noted that:

> In 2020, while the American public was focused on protecting our families from a global pandemic and helping others in need, cyber criminals took advantage of an opportunity to profit from our dependence on technology to go on an Internet crime spree. These criminals used phishing, spoofing, extortion, and various types of Internet-enabled fraud to target the most vulnerable in our society - medical workers searching for personal protective equipment, families looking for information about stimulus checks to help pay bills, and many others. (2021)

The Bureau's Internet Complaint Center (IC3) received nearly 800,000 complaints in 2020, an increase of 70 percent over the previous year. Losses from these complaints exceeded $4.1 billion. In all, there were 28,500 crimes specifically related to the virus, with "fraudsters targeting both businesses and individuals." The rapid increase in computer-enabled crime and the failure of the annual crime report to capture these trends highlights a major shortcoming of the Bureau's crime statistics.

EARLY CRIME REPORTING IN THE UNITED STATES AND THE ROLE OF EDITH ABBOTT

From the early part of the nineteenth century through the early 1930s, the collection of official crime statistics in the United States was a haphazard undertaking involving a host of local and state agencies. During this period, public officials in many large US communities were concerned about deviant behavior and embraced the notion measuring offenses as a way of addressing a pressing social problem. But nowhere was the sense of urgency about crime more pronounced than in Chicago. As Lohr notes, "In the 1900s Chicago's many newspapers, filled with stories of murders, rapes, thefts and assaults, competed to feed its reputation for criminal activity" (Lohr 2019, 137). In order to address the seeming growing criminal threat in the municipality, the City Council set out to empirically assess the extent of the crime problem in Chicago. It charged its Committee on Crime to report on the frequency of "murder, assault, burglary, robbery, theft, and like crimes in Chicago" (Lohr 2019, 138). To carry out the investigation, the City engaged the services of Edith Abbott, an eminent social worker, economist and statistician. Abbot had extensive research experience and expertise on the subject of crime in Chicago. She earned her PhD in 1905 from the University of Chicago and authored more than twenty books and articles on the social conditions experienced by women and children in rapidly-industrializing Chicago. Often overlooked as a social advocate pioneer, Abbott's statistical reports led to material improvements in Chicago's living and working conditions, and represents a critical part of her legacy.

Abbott's "Statistics Relating to Crime in Chicago," published in 1915, transcended what had been the standard for statistical summaries to that day. As Lohr notes, "she investigated the quality of the data sources, commented on how to interpret the statistics, and made numerous recommendations— still relevant today—on how to obtain better statistics about crime" (2019, 138). Following extensive research, gathering data from Chicago's criminal justice system, she concluded that existing statistics about crimes known to the police ("criminal complaints") were of dubious value, as they significantly understated the amount of crime actually being committed. Abbott's report, which included a recommendation that the police department develop a system for recording crimes in a systematic fashion, helped establish the foundation for the Uniform Crime Reporting system in the 1920s.

The need for reliable crime data at a time when it appeared offenses were escalating rapidly was apparent to legal observers and ordinary residents alike. In 1926, legendary attorney Clarence Darrow wrote that no "intelligent person can examine the statistics which are at present available and come to any satisfactory or defensible conclusion as to the number of crimes committed in the United States" (Lohr 2019, 141). The following year, members at the meeting of the International Association of Police Chiefs (IAPC), advocated for the consistent and uniform collection of crime data. According to Decker, "the Chiefs sought to reduce media pressure resulting from their reporting of sporadic crime increases, which often resulted in some police departments "cooking the books" to reduce the amount of recorded crime, even though there was no reduction in reported crime to the police" (Decker 2020, 586). At the meeting, participants developed a list of seven main crime classifications to track in order to assess fluctuations in crime. This was the beginning of what is now known as the Uniform Crime Report (UCR). In assessing the list of offenses identified by the Chiefs, the National Academy of Sciences notes:

> The problem with the list of crimes developed by the assembled police chiefs in the late 1920s is not that it is uninformative—the original Part I crimes were chosen in large part for their salience to the general public, and they remain serious events of interest today. Rather, the issues are that the list of Part I crimes have so successfully "defined"—and limited—what is commonly meant by "crime in the United States" and that the lists of both Part I and Part II crimes have remained so relatively invariant over the years. (Decker 2020, 589)

THE UNIFORM CRIME REPORT

Although it lacked authority to collect data, the Uniform Crime Report Committee of the 1920s was charged with bringing order to the government's record-keeping systems. In January 1930, 400 cities, representing about 20 percent of the US population, participated in the initial report. The report had changed little in terms of the criminal offenses Edith Abbott and the Chicago Committee examined in 1914. The same year Congress authorized the Attorney General to gather information on crime nationally, which subsequently designated the FBI to serve as the national-based agency to administer the program. Every year since, statistical information based on uniform classifications for reporting offenses have been obtained from a growing number of law enforcement agencies (nearly 20,000 in 2020). Of the UCR, the FBI states:

> The program's primary objective is to generate a reliable set of criminal statistics for use in law enforcement administration, operation, and management; however, over the years, its data have become one of the country's leading social indicators. Eight major classifications of crime, known as the crime index, are tracked to gauge fluctuations in the overall volume and rate of crime.

Between 1960 and 2004, the UCR program formally referred to its list of offenses as the Crime Index. The creators of the UCR Program chose the crimes in its index because they are serious, occur with regularity in all areas of the country, and are likely to be reported to police. According to the FBI, the UCR Program collects data about Part I offenses in order to measure the level and scope of crime occurring throughout the nation. The program divides offenses into Part I and Part II crimes. Each month, participating law enforcement agencies submit information on the number of Part I offenses they are aware of in their area. The crimes include those violations most likely to be reported to the police including criminal homicide, forcible rape, aggravated assault, burglary, robbery, larceny, motor vehicle theft, and arson. Major Part II offenses, about which there is only arrest data, include stolen property, embezzlement, forgery and counterfeiting, fraud, weapons, sex offenses, prostitution and commercialized vice, among others (FBI 2012).

MODIFYING AND EXPANDING THE UCR

There have been just a handful of changes to the original report. In 1960, the UCR program began collecting national statistics on law enforcement officers killed in the line of duty. In 1972, attacks on law enforcement officers were also included in the data collection process. Later that decade, arson was added to the UCR program as a Part I Index crime.

Hate Crime Statistics

In the Spring of 1990, President George W. Bush signed the "Hate Crime Statistics Act" into law, mandating that the Attorney General, as part of the UCR program, collect data on crimes that were committed because of the victim's race, religion, disability, sexual orientation, or ethnicity. The bill was hailed by many in the LGBTQ+ community for being the first bill of its kind to specifically identify gays, lesbians and bisexuals in crime statistics. The first annual Hate Crimes report produced by the UCR system drew upon data from eleven states that volunteered to submit their statistics to the program as a prototype. Since 1990, the Hate Crimes Report has been expanded to include crimes committed against people with physical or mental disabilities. In 2009, the Matthew Shepard and James Byrd Jr. Hate Prevention Act was passed by Congress which amended the 1990 Act to include crimes motivated by any bias against gender or gender identity. In 2013, bias crimes in the religion category were expanded to include all groups identified by the Pew Research Center and the Census Bureau.

The UCR Hate Statistics Program also began gathering data on bias crimes based on the category of race, ethnicity and ancestry in 2015. Although the program has gone well beyond its predecessors in collecting data about crimes related to prejudice against people because of their race or sexual orientation, critical observers believe the United States does a poor job of accurately tracking hate crimes (Berry and Wiggins 2018; Glickhouse 2019). It has long been recognized that official hate crime statistics understate the magnitude of the problem (Southern Poverty Law Center 2006).

In recent years, the underestimation of hate crime has drawn renewed attention. As Nakamura (2021) notes "a spate of high-profile assaults on Asian Americans has renewed long-standing criticism from Democrats and civil rights groups that the U.S. government is vastly undercounting hate crimes, a problem that they say has grown more acute amid rising white nationalism and deepening racial strife." Nakamura observes that although federal law mandates that the FBI collect data on hate crimes each year, the effort has long been plagued by incomplete and inconsistent data provided by the nation's 20,000 state, municipal and tribal law enforcement agencies. Although recent attention has focused on hate crimes against racial minorities, especially Asian Americans, official statistics on crimes against the LGBTQ+ community have also been critiqued (Hauck 2019).

In 2013, the UCR altered its conceptualization of rape, marking the first time the definition of a major criminal offense had been changed (Basu 2012; Savage 2012). The new definition was designed to be more inclusive of all forms of sexual penetration and a more accurate reflection of state criminal codes. In 2019, the FBI announced that the UCR program would begin collecting National Use-of-Force data, with goal of collecting a comprehensive view of the circumstances and officers involved in such incidents.

Owing to the significant economic impact of cargo theft, Congress passed the USA Patriot Improvement and Reauthorization Act of 2005 requiring that reports of cargo theft collected by federal, state, and local officials be reflected as a separate category in the UCR system. Publication of cargo theft data began in 2013. In 2008, Congress passed the William Wilberforce Trafficking Victims Protection Reauthorization Act, requiring the collection of human trafficking data. The Act required that distinctions be made between prostitution, assisting or promoting prostitution, and purchasing prostitution when gathering and presenting data. In 2013, human trafficking—in two forms—as commercialized sex and as involuntary servitude—were added to the Part I Index (Farrell and Reichert 2017). While identifying human trafficking as a serious offense, recent research suggests that the UCR has underestimated the extent of the problem. According to the National Institute of Justice, the labor and sex trafficking data that appears in the Uniform Crime Reporting Program may significantly understate the extent of human trafficking crimes in the United States.

CRITICISMS OF THE UCR

Since it was first created, a number of criticisms, some of which remain germane today, have been leveled at the FBI's Uniform Crime Report. Perhaps the most common criticism is that the index fails to include all crimes committed in a particular period.

Underreporting Crime

Instead of presenting data on all crimes, the UCR includes only those violations that come to the attention of law enforcement agencies who then submit criminal reports to the Bureau. In 2017, the law enforcement agencies that were active in the UCR program covered more than 310 million residents in the United States, or 98.4 percent of the total population (FBI 2018). In all, there were about 10 million people living in jurisdictions that failed to submit crime reports. The UCR also understates the true magnitude of crime due to the FBI's "hierarchy" rule. The hierarchy rule requires law enforcement agencies to report only the most serious crime when multiple offenses occur. For example, if an offender breaks into a home, steals valuables, and murders the occupants, only the homicides are reported by law enforcement to the FBI; the forced entry and theft offenses are ignored as part of the official crime statistics.

The consequences of using the UCR to produce an official crime index are socially significant for several reasons. Perhaps most importantly, the failure of the index to account for all criminal behavior means that the data cannot convey changing crime trends within the nation. Specifically, the ineffectiveness of existing crime statistics to fully record computer-enabled crime calls their veracity into question. Further, undercounting certain types of crimes disproportionately affects some groups more than others. For some time, social scientists have persuasively argued that sex-related crimes against women, including rape and domestic violence, have been significantly underestimated when the FBI's Uniform Crime Report is exclusively relied upon (Walsh and Hemmens 2018; Kimble 2018). Renewed skepticism with respect to these recording errors was directed to the Bureau's official crime statistics during the early stages of the pandemic. According to many sources, law enforcement agencies failed to capture the true magnitude of domestic violence committed against women and children during the public health crisis (Bradbury-Jones and Isham 2020; Graham-Harrison, Giuffrida, Smith, and Ford 2020). As will be detailed next, sexual violence, domestic violence, and child abuse are just some of the crimes the UCR underestimates in calculating the nation's official crime rate.

Many factors influence whether crimes committed by individuals against others are officially reported, and counted as a part of the FBI's official crime statistics. One issue is the extent to which individuals believe reporting a crime will be recorded and acted upon by authorities (Yoon 2015). Another key variable is the number of law enforcement authorities available to whom citizens can file official reports. In 2016 there was an average of 16.8 officers and 21.4 total law enforcement personnel for every 10,000 residents in communities with a population of more than 25,000. But the data also reveal a wide disparity across metropolitan areas from 71 law enforcement personnel per 10,000 residents in Atlantic City, to a low of 4.2 in Lincoln, California (Governing 2021).

The wide variation in the number of law enforcement personnel across the country creates a differential opportunity structure for residents to report crimes. The more law enforcement personnel there are for residents to contact, the more likely a crime will be reported, recorded, and submitted to the FBI. As highlighted in the Introduction, during the initial stages of the pandemic, the nation's crime rate was suppressed by social distancing measures and from local police departments temporarily overlooking minor violations (Melamed and Newall 2020).

Another variable is the relative discipline law enforcement personnel exercise in recording a reported offense. Additionally, the insurance status of stolen objects plays a role in determining whether thefts or robberies will be officially reported. An important reason new, computer related crimes are not reported rests in the fact that victims have little incentive to register a complaint with law enforcement. For example, in most cases of credit card fraud, a consumer can resolve the matter with a phone call. A physical break-in involving a comparable amount of financial harm, more often entails the completion of a formal police report.

Many social scientists have maintained that "street" crimes are heavily underreported to law enforcement. The magnitude of unknown crimes represents the "dark" figure of crime, a term first used by the Belgian sociologist Adolphe Quetelet. The dark figure of crime has been systematically studied by social scientists for decades (Biderman and Reiss Jr. 1967). The omission of criminal incidents that are not recorded in official statistics has several important negative implications for the criminal justice system and for society as a whole. As Skogan points out, "a great deal of criminal activity in America goes unrecorded, largely because it is not reported to the police. This pool of unrecorded crime has several consequences: it limits the deterrent capability of the criminal justice system, it contributes to the misallocation of police resources, it renders victims ineligible for public and private benefits, it affects insurance costs, and it helps shape the police role in society" (1977).

Periodic changes in the nation's formal laws, statutes, and regulations also complicates the official crime rate, when utilizing the Uniform Crime Reports. The UCR is an unreliable gauge of crime because activities that at one point in time were legal can later be redefined by authorities as misdemeanors or felonies. A case in point, as highlighted by social activists and investigative journalists, is the increasing tendency by local jurisdictions to criminalize homelessness (Sarma and Brand 2018; National Law Center on Poverty and Homelessness 2019). In 2019, the National Law Center on Poverty and Homelessness (NLCPH) found a 50 percent increase in the number of municipalities that prohibited sleeping citywide since 2006, and a 29 percent increase in the number that banned sleeping in particular places.

The Center also found significant increases in laws that criminalize behavior engaged in by homeless individuals, including sitting and lying down in public, loitering, loafing and vagrancy, begging, living in vehicles, food sharing and scavenging (2019). Texas' efforts were typical of recent ordinances enacted to criminalize the homeless. In the summer of 2021, the state outlawed camping in public spaces; individuals charged with camping outside of designated areas now face fines of up to $500.

Clearly, the heightened presence of homeless individuals in metropolitan areas is not synonymous with an increase in criminal behavior, but changes in the legal code make it appear that way. Conversely, the decriminalization of cannabis consumption in many states may create the misimpression that deviant behavior is declining, when it is merely states' regulatory regimes that are changing (Maier et al. 2017). By mid-2021, twenty-seven states and the District of Columbia had legalized or decriminalized small amounts of marijuana (Hartman 2021).

White-Collar Crime Statistics

In addition to underestimating sexual and domestic violence, the UCR has been criticized for understating the magnitude of white-collar and corporate crime (Cohen 2013; AIT 2015; Stewart 2015). The concept of white-collar crime was first introduced by Edwin H. Sutherland at the annual meeting of the American Sociological Society in 1939. In his presidential address, Sutherland raised concerns over the preoccupation with low status offenders to the neglect of offenses committed by those in higher status occupations. He observed that: "crime statistics unequivocally show that crime as popularly conceived and officially measured has a high incidence in the lower class and a low incidence in the higher class, less than 2 percent of the persons committed to prison in a year belong to the upper class" (1940). Sutherland argued that theories of deviant behavior based on crime statistics and case histories were misleading and incorrect.

Although white-collar crime can be defined in terms of the offender, the offense, or the organizational culture in which the crime is committed, the FBI has opted to adopt the approach that emphasizes the criminal offense. Barnett notes that the Bureau has defined white-collar crime as "those illegal acts which are characterized by deceit, concealment, or violation of trust and which are not dependent upon the application or threat of physical force or violence" (2000). Some have criticized this conceptualization of white-collar crime because it emphasizes the offense rather than the background of the offender.

Critically, Burke and colleagues note that corporate crime inflicts far more damage on society than all street crime combined, whether measured by death, injury, or dollars lost (2019). The FBI estimates that white-collar crime costs between $300 and $600 billion annually, a figure that far surpasses the cost of losses due to property crimes and bank robberies combined (AIT 2015). Although the UCR excludes costly white-collar crimes such as stock market fraud, tax evasion, and hazardous waste dumping, it does provide some data on the extent of corporate crime. The FBI's "Financial Crimes Report" includes information on crimes such as insider trading, fraud schemes, and medical fraud. The Bureau's most recent Financial Crimes Report reviewed the Agency's investigations for the years 2007–2011. During 2011 the Bureau examined 242 corporate fraud cases that resulted in 241 convictions of corporate criminals, $2.4 billion in restitution orders and $16.1 million in fines from corporate criminals (FBI 2012).

Internet Crime Statistics

Another important criticism leveled at the UCR is its ineffectiveness in tracking the way criminal behavior is evolving. The Bureau acknowledges that "it is well documented that the major limitation of the traditional Summary Reporting System is its failure to keep up with the changing face of crime and criminal activity. The inability to grasp the extent of white-collar crime is a specific example of that larger limitation" (Barnett 2000). Cybercrime is a noteworthy example of a rapidly growing transgression that inflicts immeasurable harm but goes largely uncounted in official crime statistics. Cyberattacks are the nation's fastest growing type of crime (Decker 2020). Independent studies show a nearly 500 percent increase in healthcare email fraud in recent years, a significant growth in online crimes against children, a rapid advance in cyberattacks through mobile devices, and a dramatic rise in global phishing attacks. The financial impact of ransomware attacks are substantial. Estimated to cost more than $300 million in 2015, their impact exceeded $5 billion by 2017. By 2016, cybercrime was costing the US economy between $57 and $109 billion annually (Council of Economic Advisers 2018).

In May 2000, the FBI created the Internet Fraud Complaint Center also known as IC3, to provide the public with a reliable and convenient way to submit information concerning suspected Internet-facilitated criminal activity. In its first full year of operation, the IC3 logged 49,711 complaints, most involving Internet auction fraud, nondelivery scams, and the infamous West African letter scam. In 2020, the IC3 received more than 790,000 complaints from the public, with reported losses exceeding $4.1 billion. The voluntarily submitted complaints are the basis for official statistics on cybercrime. Despite the growth and magnitude of this type of crime, it is estimated that the Bureau's online database captures only about 12 percent of cybercrime.

The growth in offenses committed online and the inability of official statistics to capture these incidents may be obscuring trends within the nation's overall crime picture. Increasingly, official crime statistics fail to shine light on key developments occurring within the county. The FBI reports there were 4,251 bank robberies in 2016—a decline of 45 percent compared with 2004. At the same time, there were nearly 300,000 reports of people being victimized from crimes committed on the Internet (Wexler 2018). In a notorious case, thieves working with computer experts stole $45 million from thousands of automated teller machines in a ten-hour period (Marzulli 2013). The loss from this one case exceeded the total losses from physical robberies of banks over an entire year. Clearly, the existing statistical series fails to account for these types of important changes in the way criminal offenses are being commissioned. Indeed, only an estimated 15 percent of the nation's fraud victims report their crimes to law enforcement. Although estimates are imprecise, it is plain that millions of people are victims of Internet crimes each year.

As Decker explains, the lack of reliable data on cybercrime complicates the ability of authorities to track and prosecute individuals involved in the activity:

> Experience demonstrates that crime data can successfully be used to counter and address criminal trends and to effectively train and deploy law enforcement

officers in the areas where they are most needed. Absent data that informs cyber-crime-fighting decisions, policymakers and criminal justice leaders cannot appropriately respond to this prolific crime.

The inability of official statistics to effectively measure the growing importance of cybercrime underscores the way the data can be misused. Politicians, law enforcement professionals, and policy analysts rely upon the Bureau's statistics as a tool to address crime. But using the current statistical series certainly invites problems as they are unable to present comprehensive information on the evolving face of computer-enabled crime in America. (2020)

Emphasizing Street Crime

The disparity between the social costs that accrue from "suite crime" and computer-enabled crime versus those entailed from street crime highlights the social significance of official statistics and why they deserve sustained critical scrutiny. Many Americans, conditioned by years of mainstream news coverage, believe the typical criminal is a poor minority youth living in the inner-city. The notion that an older white male, residing in a comfortable suburban neighborhood could just as accurately be viewed as a typical criminal seldom occurs to consumers of the Bureau's annual Crime in the US report.

The emphasis on street crime in the UCR illustrates the way official statistics guide our attention about important social issues in certain directions and away from others. Official government statistics on crime that underestimate costly criminal activities undermine the nation's economy and its social structure. Moreover, the relative inattention paid by the justice system and the mass media in addressing white-collar, corporate and computer-enabled crime leads the general public to misinterpret the broader scope of the nation's crime problem. For all the angst that frequent reminders of the street crime problem may create, most have little idea that the official figures only scrape the surface of the larger crime picture. Finally, the UCR's focus on street crime has immediate political implications in reducing the likelihood of punishment among certain classes of criminal offenders, specifically, white-collar offenders whose crimes are more likely to go unreported or underreported than those found guilty of street crimes (Olejarz 2016; Bonn 2017). If the FBI and mass media devoted as much attention to white-collar crime as they do to street crime, Americans would perceive an equal need to punish the privileged perpetrators of those offenses.

ALTERNATIVES TO THE UCR

As with the official unemployment rate, and life expectancy statistics, it is important to consider alternative statistics on crime. Of particular interest are the National Crime Victimization Survey (NCVS) and the National Incident-Based Reporting System (NIBRS). The NCVS began as a Census Bureau pilot program in 1971. At that time, the Bureau integrated questions related to victimization into its existing "Quarterly Household Survey." In 1991, after extensive bureaucratic wrangling within the Department of Justice, the ongoing effort to document the extent of victimization in the United States became the responsibility of the Bureau of Justice Statistics (BJS). Subsequently, the BJS renamed the series the National Crime Victimization Survey. The NCVS, which includes crimes that go unreported in the UCR has been identified as a "better method of collecting crime statistics" (Feagin 1997) and an important step in understanding the true magnitude of criminal victimization in America. The NCVS has been particularly helpful with regard to crimes that disproportionately affect women and marginalized groups.

In creating the NCVS, additional questions concerning sexual assault covering a broader range of incidents, such as verbal threats of rape and unwanted sexual contact involving threats to the victim, were included. Further, the NCVS, unlike the UCR, records data on rapes against both sexes. In addition to altering and expanding criminal categories, the BJS modified the wording on survey questions in an effort to minimize anxiety survey respondents may experience when reporting a sexual crime. Over time, the BJS also added a series of questions about hate crimes to shed light on race-based offenses that are neglected in the UCR.

Although the NCVS has done much to shed light on the dark figure of crime in America, particularly regarding crimes against women, it too has been criticized as an accurate barometer of crime in the United States. One commonly expressed critique of the NCVS its that it excludes homicide and thus neglects the most serious type of victimization one could experience. Additionally, the NCVS is a household survey and thus tends to understate the extent of victimization among businesses in the form of pilferage, burglaries, and vandalism. It has also been criticized because it relies entirely on the memory and veracity of the survey respondents.

The National Incident-Based Reporting System (NIBRS)

Presently, the UCR program is transitioning from the system it has relied on since 1930 to a new "National Incident-Based Reporting System" (NIBRS).

The UCR's Summary Reporting System was formally retired on January 1, 2021. In addition to tabulating offenses, the new data collection system is designed to gather and document information about each criminal incident, including the characteristics and relationships of victims and offenders, crime locations, weapon use, and the value of property lost. Other information, such as whether the crime was cleared is also included to help embed the "incident" in a more meaningful context. In contrast to the UCR, the NIBRS does not have a hierarchy rule, so when multiple offenses occur in the commission of a crime, they are all recorded, not just the most serious one. In this way, the NIBRS addresses one of the more serious criticisms that have been leveled at the UCR.

The BJS first began developing the NIBRS in the 1980s to more accurately track the magnitude of crime. The program was intended to enhance the quantity, quality and timeliness of crime data and to improve the methodologies involved in the production and dissemination of official crime statistics. In 2013, the NIBRS system began collecting data on each incident and each arrest within fifty-eight offense categories. Two years later, nearly 7,000 agencies or about one-third of the those currently providing data to the UCR program were participating in NIBRS. President Obama's Task Force on 21st Century Policing encouraged participation in the program, finding that greater acceptance of it could benefit policing practices and research. By 2018, the FBI reported that 44 percent of law enforcement agencies were NIBRS-compliant and that thousands of others had confirmed they would be conforming to the new system by the start of 2021.

Criticisms of the NIBRS

As with the UCR program, the new NIBRS has come under critical scrutiny. For instance, like the UCR, the NIBRS does not include all crime, only those offenses that are reported to a law enforcement agency. As a result, the "dark" or "shadow" figure of crime continues to be an issue with the NIBRS. Another important criticism aimed at the NIBRS is that the tension that exists between the program's participants interferes with the ability of the program to effectively execute its mission. The NIBRS is administered by a federal agency, but relies on the cooperation of thousands of local and state offices to provide data. In part, the issue is logistical. Local agencies with a myriad of data collection methodologies are being asked to conform to a single way of gathering statistics according to nationally-orchestrated guidelines.

More importantly, states providing data often have social policy agendas that are at odds with federal agencies. For example, the BJS is largely interested in generating empirically replicable reports that accurately reflect national fluctuations in crime, down to the finest detail. For local authorities, the purpose of gathering and circulating crime data is more immediate and palpable. For them, crime statistics provide a basis to allocate and direct law enforcement resources in their communities. Finally, some critics maintain that although the current UCR system is flawed, it nonetheless provides researchers with an accessible source of summary-based information. In contrast, the new NIBRS, with its more nuanced, incident-based approach may prove to be prohibitively cumbersome compared with the more user-friendly UCR annual reports.

CONCLUSION

Official statistics on crime warrant sustained scrutiny because they serve to create a social reality that powerful interests can manipulate for their own ends. Indeed, official crime statistics matter because they are frequently "weaponized" to achieve specific political objectives. As this critical narrative implies, crime is not an objective reality. Through the creation of both Part I and Part II UCR statistical series, the FBI tacitly acknowledges that the concept of "crime" is a fluid one, subject to changing interpretations. The Department of Justice's National Crime Victimization Survey (NCVS) further underscores the fact that no one "official" measure of crime captures the true extent of all deviant behavior.

Moreover, official crime statistics can be politicized in profound ways to persuade Americans to justify the adoption of policies they otherwise would not favor, such as enhanced budgets for additional law enforcement personnel. In recent years, government generated crime statistics have been utilized in the ongoing struggle between gun rights advocates and those seeking to impose restrictive regulations on the sale and use of guns. In the wake of data showing a growing frequency of mass shootings in the country, it is of interest to note that official crime statistics have been used both by those calling for stricter reform (Frosch and Elinson 2021), and by second amendment advocates (Poliquin 2021). In a similar vein, official crime statistics are frequently politicized by those with broader ideological intentions. For example, the bias of official statistics toward street crime can be distorted and exaggerated to make Americans more receptive to accepting authoritarian policies in the name of greater security.

Further, official statistics can provide political cover for those wishing to take credit for decreases in crime in their communities. Over the decades, most forms of serious street and violent crime have officially declined, but overall victimizations have not subsided as perpetrators have migrated their criminal activities. As official crime reports become increasingly divorced from empirical reality with the spread of cybercrime and other forms of computer-enabled crime, political actors can shelter their claims behind out-dated, dubious statistics.

Moreover, official crime statistics matter because their narrow focus means offenses can easily be misconstrued as an accurate gauge of all crime in the county. Powerful actors, including elected officials and corporate executives have a vested interest in making the FBI's annual report appear as if it captured the nation's entire crime picture. White collar criminals benefit from the attention devoted to perpetrators of street crimes. Classifying white-collar crimes in the Part II category, despite their greater social costs, influences the way those offenses are perceived by the public and by policymakers.

Further, official crime statistics are socially significant because they have a pivotal influence in creating social policies that favor the positions of the powerful in society. Official crime statistics direct our attention to offenses powerful interests want society to focus upon, and not the entire spectrum of deviant acts that fall outside the formal legal structure. As a case in point, the problem of homelessness has been a persistent issue in the United States since the nation became an urban, industrializing power. But only recently has homelessness been officially conceptualized and counted as a type of criminal behavior. In turn, the new, and artificially manufactured homeless crime statistics can easily be manipulated by actors who wish to appeal to those opposed to publicly-financed assistance for the dispossessed.

Importantly, official crime statistics matter because they fail to provide any insight into the more general issue of deviance in society and the more specific issue of criminal behavior. Whether people are more or less willing to accept and follow well-established social norms is not a pressing matter concerning government econometricians. The concerted emphasis on street crime as represented in official statistics provides little, if any, indication about whether society is becoming more socially cohesive and integrated, or more anomic in character.

In sum, as with the other government generated statistics discussed in this monograph, official crime data are socially significant and warrant sustained scrutiny because they are misused, misinterpreted and politicized. In the same way that national account statistics (GDP) are misused as the broadest barometer of a society's well-being (as discussed in the final chapter), crime statistics are misused and misinterpreted as being an accurate barometer of the extent of crime in society. The neglect of white-collar, corporate, and computer-enabled crime leads to a misinterpretation about the demographics of criminals in society, serving to generate stereotypes about offenders.

Chapter Four

The Census Bureau and the Decennial Population Undercount

This chapter on official statistics informs readers about the Bureau of the Census' efforts to provide an accurate portrait of the American population. As with the earliest endeavors to calculate joblessness, the first attempts to survey the US population began at the state level. Since 1790, a national census has been undertaken every ten years to enumerate the number of inhabitants in the United States and their demographic characteristics. The earliest version of the national census, in addition to asking the name of the head of the household, posed just five questions. The first legislation calling for a national census bureau was passed in 1840, with the United States Bureau of the Census (as it is formally known), being created in 1903 as a part of the Commerce and Labor Department. From its inception, the Bureau has experienced varying degrees of success in its efforts to canvass the entire US population and to disseminate an increasingly expansive range of social and economic data.

As with the others in this volume, this chapter highlights the ways official statistics are misused, misrepresented, and politicized. Official data collected by the Census Bureau are misused when they become the basis for mis-distributing funds to communities. The Bureau's official statistics are misinterpreted because the public largely accepts uncritically what the government reports about the population. The minority undercount, for example, creates the impression that the nation is less "brown" than is otherwise the case. Finally, the entire census enumeration process has been politicized, not just the numbers themselves. For instance, in advance of the 2020 canvass, President Trump insisted the census include a citizenship question. Had this proposal been embraced, it would have suppressed the official count among minorities, immigrants, and undocumented workers.

Further, the inaction of some states in "getting out the count" was part of a broader agenda designed to minimize public resources going to minority communities. The negligence shown by conservative interests in obtaining a full count was also motivated by a desire to disenfranchise minority groups from the political process. The enumeration process was further politicized when the Trump administration called for an early end to the canvass. When that failed, attempts were made to have a second set of data prepared that would have allowed for the exclusion of noncitizens in the final count. Finally, the process was politicized when the Bureau's director was forced to resign a year before his term ended amid charges he was pressuring government employees to produce an alternative set of census statistics for apportionment purposes.

Even before these events and the pandemic complicated the canvass, the 2020 census faced several logistical challenges. These included the prospect of an unprecedented citizenship question as well as underfunded preparation. The introduction of new, experimental methodologies, such as relying on Internet responses, and using administrative records from federal agencies, also vexed the survey. Before the pandemic's onset, the Bureau estimated it would cost the federal government nearly $16 billion to fully enumerate the population. This figure is 25 percent greater than the Bureau's initial projection and three times larger than what government spent to count the population in 2000. Despite the outlay, the Government Accountability Office (GAO) placed the 2020 census on its list of "high-risk" government projects (Clark 2019). According to the GAO, an estimated funding shortfall of $3.3 billion and leadership gaps left the US Census Bureau without the resources it needed to prepare for the 2020 census and to ensure a complete enumeration of the US population. Taken together, these factors significantly heightened the economic and political impact of the bureau's numbers.

THE 2020 CENSUS

As the canvass began to unfold, many states devoted substantial resources toward making the decennial census an accurate depiction of the US population. Generally, the states that made the greatest efforts were ones led by Democrats. In states where the GOP was the dominant party, comparatively little was spent on get-out-the-count campaigns. Republican officeholders, and conservatives generally, worried that emphasizing the economic impact of the census would energize minorities to become more active in the electoral process. With minorities being far less likely to endorse candidates from the Republican party, there was little incentive for GOP leaders to obtain a reliable count. In the midst of the enumeration effort, the pandemic complicated matters by interfering with the ability of canvassers to reach interview respondents using the traditional methodology of sending field workers to homes. As Levine notes, the Bureau was already a fragile operation that faced immense challenges counting minority populations and other groups in the United States before the outbreak of a global pandemic (2020).

In particular, the pandemic heightened the inability of the census to reach traditionally undercounted populations. The *PBS News Hour* reported in early 2020 that the COVID-19 outbreak could hinder many people in local and state communities in the United States through the loss of federal funding for years to come (Khan 2020). Wines (2020) and Levine (2020) noted how the severe virus-related limitations on personal contact upended a decade of planning on the Bureau's part to accurately record hard-to-count populations. The Bureau's inability to reach marginalized groups meant they would continue to be underserved by political representatives and public resources, perpetuating institutional racism (Capps 2020).

Get Out the Count Efforts

With congressional seats and billions of dollars in federal aid at stake, many states were eager to budget substantial sums to encourage residents to complete the 2020 census. California invested more than $185 million in "get out the count efforts" to ensure they would not lose one of their fifty-three Congressional seats and to protect the federal aid it receives (Castillo 2019). About 72 percent of all Californians belong to one or more hard-to-count groups, including noncitizens, renters, children, African Americans, Latinos, and Native Americans. The sum was unprecedented for the state. In 2000, the California legislature budgeted $25 million for the purpose of obtaining an accurate census count, but in 2010, the state allocated just $2 million as it was still recovering from the impact of the Great Recession.

California, with its large immigrant population, has a heightened incentive to ensure an accurate count given the Trump administration's early efforts to include a citizenship question on the census (Finch 2020). "At the end of the day, it's about money and power," said Diana Crofts-Pelayo, communications chief for the California Complete Count Committee, an advisory panel that coordinated education and outreach efforts for the state. "This is oftentimes more important than voting." Crofts-Pelayo said cities and states have always had an involved role in the census, but this year the count turned into a partisan issue over the proposed citizenship question. "If there is an undercount . . . it will cost the state $1,000 every year [per person undercounted] for the next 10 years," she said. The Central Valley stands out as the hard-to-count region in the state. Officials and activists are concerned Los Angeles, Monterey, and Imperial Counties also face serious challenges in obtaining a complete count.

California was not alone in spending a large sum to keep and gain political and economic resources. According to the National Conference of State

Legislatures (NCSL 2019), more than twenty states devoted at least $10 million in efforts to ensure an accurate count for their state. Illinois spent $30 million, New York allocated $20 million (NBC NewYork 2020) and Washington State budgeted $15 million in an effort to ensure they would receive fair political representation and their share of federal subsidies. In most cases, the investment would seem a prudent one to make. According to a George Washington University (GWU) study, in 2016, New York state received more than $73 billion in census-connected funds (Reamer 2018).

Despite the political and fiscal consequences, some states chose not to devote money to mobilize efforts for the 2020 decennial census. For example, neither Oklahoma nor Texas allocated money for the purpose of obtaining an accurate census count. Joe Dorman, a former Oklahoma state representative and the CEO of the Oklahoma Institute for Child Advocacy, noted the state has a history of undercounts and that a shortfall from the federal government would likely result in local tax increases for the state's residents. In Oklahoma, each person counted brings in an estimated $1,600 for ten years.

Texas' failure to fund a "get out the count effort" was particularly problematic given its large and growing Latino and immigrant populations. According to Johnson and Williams (2019), a 1 percent undercount in the state could cost it $300 million annually for the next decade in federal funds. Texas' difficulty in obtaining an accurate count is compounded by the fact roughly one-quarter of the state's residents live in hard-to-count neighborhoods, partly because many of Texas' rural areas lack access to high-speed Internet. According to projected census data, demographic shifts could yield the state three additional House seats and hundreds of millions in aid but due to the lack of a coordinated effort to obtain an accurate count, the state may fall short.

According to the GWU report, Miami-Dade county ranked first in the nation in terms of risking an undercount in the 2020 census (Reamer 2018). According to Florida state politicians, the large, growing undocumented population in the state, plus crackdowns on sanctuary city measures were discouraging local residents from interacting with anyone representing the government. After the 2010 census, the state lost an estimated $20 billion in federal funding between 2010 and 2020 because hard-to-count populations in Florida were not included in the official statistics. Overall, Florida had the third largest undercount in the nation, with an estimated 1.4 million persons left uncounted (Miami Herald Editorial Board 2019). The George Washington University report identified Maricopa County (Eltohamy 2019) as being at risk of having the second greatest undercount in the 2020 census—which could end in losses of up to $13.2 million in federal funding for some underserved communities in the County. The report projects an undercount of 70,500 in the county, which represents 1.3 percent of the 2020 estimated population of 4.5 million.

Unsurprisingly, black, Latinx, and undocumented communities in the county are the most likely to be undercounted, the report revealed. The federally funded programs that are most at stake include Medicare, education grants, and health insurance for lower income children (Centers for Medicare and Medicaid Services). The loss of funding for these programs, particularly those that service low-income minority residents in Maricopa County, vividly demonstrates how official census statistics can play a pivotal role in perpetuating systemic racism in the United States. Decade after decade of undercounting by the Census Bureau leads to an inestimable degree of harm to individuals, neighborhoods and communities of color by denying them basic social services and the political representation needed to advocate for them.

A 2018 report from GWU identified thirty-seven states that would have lost millions in federal funding in fiscal year 2015 if their populations had been undercounted by 1 percent during the 2010 census. Texas would have been shortchanged by $292 million, Pennsylvania by $222 million, Florida by $178 million, Ohio by $139 million, Illinois by $122 million, and Michigan by $94 million (Reamer 2020). Given the potential for losing political representation and federal funding, the 2020 census incentivized numerous cities to become engaged in "get out the count efforts" that previously had not done so. For example, New York budgeted $40 million (Slattery 2019). In states without a budget for the census count, such as Texas, cities like Dallas-Fort Worth and El Paso coordinated and funded their own get out the count efforts.

OVERVIEW: THE SOCIAL IMPLICATIONS
OF THE DECENNIAL CENSUS

The economic and political implications of the decennial census are myriad in scope and immense in their magnitude. First and foremost, official census statistics are used to allocate seats in the US House of Representatives, and by extension, votes in the Electoral College. Each census sees local, state and federal legislative district lines redrawn in wake of the official count. Researchers have tried to quantify the national impact of an undercount on Congressional apportionment. In one study, (Seeskin and Spencer 2015) found that between ten and fifteen House seats would go to the wrong state if the average percentage error for state population data was 4 percent. Seeskin and Spencer also looked at how an inaccurate census would affect the distribution of funds for 140 federal programs that rely on census data to determine eligibility. According to these authors, a 4 percent error in the 2020 census would result in an estimated $60 to $80 billion in federal grant money being misallocated over the following ten years.

Additionally, the census drives the distribution of $1.5 trillion in tax dollars for communities nationwide and vital public needs: schools, housing, mass transportation, health programs, infrastructure (roads, highways, bridges), public safety, Medicaid, emergency services, and more. As the Bureau's spokesperson states, "billions in federal dollars flow to state governments to the local level. It's about power and money. It shapes the future" (Vox 2020). According to the Bureau, the count determines how $675 billion is spent annually. A report by George Washington University estimates that number may actually be more than $1.5 trillion.

The degree of political attention devoted to official census numbers is unique among the statistics covered in this book. Following the 2010 census, more than two hundred jurisdictions around the country legally challenged federal census figures, alleging an undercount of some form.

The Social Implications of the Decennial Census: Solidifying Systemic Racism

The government's consistent underestimation of minorities in official statistics is an important theme in this book, and of particular importance in this chapter. The Census Bureau's ineptitude in fully counting minority populations contributes to systemic racism, one of the more critical issues confronting communities of color today. Systemic racism (Carmichael and Hamilton 1967) refers to discriminatory practices that have been normalized within the major institutions—criminal justice, education, housing and employment—of US society.

The data derived from the decennial census has important social implications that serve to reproduce institutionalized racism. As noted, the data are used to allocate seats to the House of Representatives and to determine the distribution of more than a trillion dollars in federal funding for a variety of community-based social and physical infrastructure programs. Clearly, undercounting minority communities that have historically been underserved only serves to exacerbate institutionalized discriminatory practices.

Moreover, the Census Bureau's statistical portrait reinforces systemic racism in a symbolic way. The undercount among people of color suggests minority Americans are undervalued compared with white Americans and begs the question: "Black Lives Matter," but do they count? From the perspective of the statistical agency responsible for quantifying the entire US population, the answer appears to be "no," or perhaps, "yes, but not as much as others." Lastly, although the Census Bureau discloses that the United States is more diverse than ever (US Census 2021), the decennial undercount of minorities yields a distorted demographic image of society as a whole, one that is considerably whiter than its actual composition.

The Census Bureau's inability to enumerate the entire population is reflected in the publication of an "official" undercount the Bureau issues in the aftermath of the decennial census. Historically, it has been the most marginalized groups in US society that have been left out of the official count.

The Social Implications of the Decennial Census: The Case of Detroit's Water Crisis

When a community is undercounted, it receives fewer federal dollars. When a community is overcounted, an inordinate number of federal dollars are distributed there. The misallocation of nearly a trillion dollars of funding further exacerbates inequities across the country. In communities where residents rely on federally funded programs, reduced spending can have a huge impact, putting more pressure on social service nonprofits to meet needs that were once fulfilled through federal expenditures. Beyond serving as a method for allocating funding for social programs, population data is also crucial for economic development. Businesses use this information to determine where to build new stores or factories—decisions that can create jobs and improve the local economy. Undercounting can also affect important political representation. Depending on the extent of an undercount, a state may lose seats in the House of Representatives.

The impact of a census undercount on Detroit's water crisis will be examined to illustrate the social implications of using official statistics. As McLean (2019) documents, in 2017, four years after the beginning of the water crisis in Flint, public health officials found dangerously high levels of lead in more than 1,600 children aged six and under. As many recognize, lead poisoning has long been implicated in causing learning difficulties, seizures, and developmental delays in children. In 2017, Flint applied for a $1.34 million CDC grant that would have enabled it to hire more staff to focus on preventing childhood lead poisoning. The grant would have allowed city officials to increase both the quantity and quality of testing of the most at-risk kids. But it was denied. According to the CDC, Detroit's population failed to reach the minimum threshold of 750,000 people required to meet eligibility requirements for its grants. The 2010 US census calculated Detroit's population at 713,777. Although it is uncertain Detroit's population was undercounted by the roughly 36,000 needed to qualify for the grant, there is evidence that Detroit's population in 2010 would have reached 750,000 with a more accurate count.

Detroit's lost opportunity underscores the importance of obtaining and using accurate official statistics, and in this case, once again underscores the ties between official statistics and systemic racism. For residents of Detroit, the underenumeration of children under age five in the 2010 census was a key factor preventing local officials from receiving aid. According to the Census Bureau's analysis in 2014, nearly 1 million children—4.6 percent of all children under the age of five in the United States—were not represented in the 2010 count.

As documented in greater detail below, children who are Latinx or black were undercounted at higher rates than white children. According to the Census Bureau, undercounts typically reflect the experience of children who have complex, nontraditional living arrangements. This includes children who split time being with parents who do not live together, or those from hardest-to-count families, such as ones who live in high-poverty neighborhoods or in rental housing. Detroit's population of children under five is higher than the national average and according to research conducted by the City University of New York, several of Detroit's neighborhoods are among the hard-to-count in the country.

Hard-to-count communities often include young children, racial and ethnic minorities, non-English speakers, low-income people, disabled individuals, people experiencing homelessness, and those living in nonstandard housing. According to the census, Detroit has a poverty rate of 38 percent, 85 percent of its population are considered ethnic minorities, more than 10 percent of its population uses a language other than English at home, and 20 percent of its population is disabled. To further complicate matters, one in five Detroit residents is evicted each year, a problem that inordinately affects women of color, further adding to an undercount in the area. Lastly, the Great Recession, which severely damaged the city's economy, likely played a part in undercounting Detroit's population. In brief, although many factors were at play, Flint's inability to effectively deal with its water crisis had a disproportionate impact on racial and ethnic minorities. The health-related ripple effects of the water crisis may hamper the life quality of Flint's heavily minority population for decades into the future. Detroit is a textbook case of a hard-to-count area.

According to Kurt Metzger, a former Michigan Mayor and entrepreneur who created a local data gathering organization known as Data Driven Detroit, city leaders in 2010 were preoccupied with Detroit's high unemployment and foreclosure crises, not with obtaining an accurate census count. He expected an undercount, but the end result was more severe than anticipated. "While I have no exact undercount in mind, said Metzger, "I was floored when I heard the 2010 count. I knew there was going to be a significant population loss even without an undercount, but was expecting something closer to 775,000." Unlike many in the metro area, Metzger understood that the census undercount was responsible for the city being disqualified from receiving the CDC grant. As this example signifies, a relatively small miscount can lead to the loss of a Government social program that potentially had benefits for local children, the vast majority of whom were from minority backgrounds.

According to Detroit officials, the 2017 grant would have increased by one-fifth the number of children under six who were being tested for lead, allowed the city to collect better data to identify higher-risk populations, and enabled them to better identify children who have been exposed so they could be networked with appropriate social services. According to Metzger, "many Detroiters had no interest in being counted and the city never worked to convince them otherwise" (McLean 2019). In short, given the multiple, overlapping factors at play, it is reasonable to surmise that Detroit's 2010 undercount was larger than the national average and that under different circumstances, the city would have qualified to receive an important grant to help an at-risk population. Instead, the official statistics gathered by the census will continue to play a role in indefinitely perpetuating institutionalized racism in the form of health care disparities for the residents of Flint.

Detroit residents and their elected representatives were not the only ones with concerns after the 2010 census. Elected officials and social activists openly worried about the accuracy of the count and its implications for their communities. In Syracuse, Michael Collins, executive director of the Northeast Community Center, noted that the prospect of a citizenship question increased distrust in the census process among local immigrant and refugee communities. His Center worked tirelessly to convince local residents it was both safe and important to participate in the census. According to the "Daily Orange," the census has historically undercounted the population of Syracuse. In Los Angeles, local officials highlighted a study that showed in a worst-case scenario, the County would lose $586 million annually in federal funding due to an undercount (Torralba 2020).

THE CENSUS' ONGOING UNDERCOUNT OF THE US POPULATION

The Census Bureau recognizes there are a number of demographic groups they fail to fully enumerate during the decennial census canvassing process. These "hard-to-locate, hard-to-contact, hard-to-persuade, and/or hard-to-interview" would-be respondents include children under the age of five and people who live in unconventional households, such as multi-generation or blended families. Racial, cultural and linguistic minorities, and immigrants also feature prominently among hard-to-count groups. Similarly, lesbian, gay, bisexual, transgender or other persons with a non-traditional sexual orientation or form of expression tend to be difficult to fully enumerate. Likewise, those with physical or emotional disabilities along with the homeless and others living in nontraditional housing are hard-to-count. Finally, lesser educated, and low-income persons people who live in remote, geographically isolated rural areas, and those who reside in difficult to access buildings or who live in strictly-enforced gated communities are among the groups the census undercounts.

Factors Influencing Census Undercounts

Demographer, and former head of the children's advocacy group Kids Count, William O'Hare (2019) has extensively examined some of the reasons the census fails to fully enumerate the US population. In part, O'Hare attributes the census-taking process itself, including processing errors and the design of the questionnaire, for part of the undercount problem. The process can create confusion about who is considered part of a household, particularly among residents who are unfamiliar with concepts such as "household." In some cases, there might not be enough space on the form to list every person living in large households. "The Census Bureau's data collection methods have not kept pace with the rapidly changing American family," O'Hare writes. Other common reasons for census undercounts stem from people who do not respond to enumerators because they wish to conceal themselves from the federal government. Others do not trust the Census Bureau to treat their data confidentially. Still other respondents are highly mobile and may not receive a census questionnaire.

Compared with more affluent respondents, poor residents, the unemployed, and the less well-educated are less likely to understand the importance of participating in the census. Additionally, residents might be missed if their addresses are not included on the federal government's Master Address file. Examples of such addresses include illegal apartments within a multiunit structure and homes far removed from roadways. In reviewing the broad picture, O'Hare states, "perhaps the most fundamental conclusion is that there are many different reasons why people are missed in the census" (O'Hare 2019).

In addition to the groups that have traditionally been un- or undercounted, two other groups will likely experience a higher-than-average undercount in the 2020 census—those with an underlying distrust of the government, and those who lack access to high-speed Internet service. Layered on top of the problem of accessing hard-to-count groups in 2020 was an unprecedented public health crisis, a sharp economic recession, and social unrest seldom seen in contemporary US society. Any one of these factors would have had a depressing impact on the Census Bureau's efforts to fully enumerate the US population. Taken together, they all but guaranteed a 2020 census undercount.

Census undercounts are not new. Overall, the 2010 census missed an estimated 16 million people. Vermont, West Virginia, Oklahoma, and Texas were among the states with the largest net undercounts in 2010. Apart from children, minorities were particularly likely to be overlooked in the 2010 decennial census. As Mule (2012) notes, undercounts among non-Hispanic blacks have been well documented in each of the last several censuses. According to the Census Bureau, black residents were undercounted by 2.07 percent in the 2010 census, Hispanics by 1.54 percent, and Native Americans on reservations were undercounted by 4.9 percent. The Census Bureau estimates it also missed 8.5 percent of all renters in the United States in the process of calculating the population for the 2010 decennial census (US Bureau of the Census 2012).

Undercounting Children

According to O'Hare (2015), the census has a long history of undercounting children, especially young ones. The 2010 census resulted in a net undercount of 1.7 percent for US residents under age eighteen, which translates into 1.3 million children—more than three-quarters of whom were four years of age or younger. The undercount of children varies by race and ethnicity. The net undercount rate for black children aged four and younger was 4.6 percent, while the net undercount rate for Hispanic children was 7.5 percent. As O'Hare notes, "young Black and Hispanic children account for about two-thirds of the net undercount in this age group even though they only account for about 40% of the population."

One explanation offered to explain the undercount among children is that household members often are listed from oldest to youngest. As Scott reports, the undercount of young children is not a new issue: "Demographers have found undercounts of young children in the 1940 and 1950 census, and even as far back as 1880" (Scott 2020). The undercount among young children has been increasing since 1980—a period in which the official tally has been improving for other age groups. If children are un- or undercounted, then the communities in which they live will be denied equitable access to programs designed to help them lead healthy lives, such as SNAP, Medicaid, and the National School Lunch Program. Needless to say, the absence of these programs in disadvantaged communities serves to further solidify systemic racism.

Children under the age of five face a projected undercount in the 2020 census ranging from 943,000 to 1.3 million, representing 4.6 to 6.3 percent of their total population. US census officials have sought to understand why young children are more likely to be undercounted in the nation's decennial census than any other age group. In a recent study, Walejko et al. (2019) analyzed data from the federal government's 2020 "Census Barriers, Attitudes and Motivators Study" involving a national survey of 50,000 households. In the study, more than 17,000 households responded to an examination of the public's attitudes regarding the census and plans for participation. Researchers found that households with young children—those age five and younger—were less likely to be familiar with, and to complete the census form than households without young children. They also were less likely to believe their participation mattered or that determining congressional representation was an important use of census data.

Among households with young children, 60 percent of respondents indicated they would complete the census, compared with 68 percent of households without young children. The researchers also discovered that higher-income, better educated families with young children differ in many ways compared with lower-income, less educated ones. For example, "only about 18% of respondents in households with young children, with incomes of $75,000 or more, were extremely concerned or very concerned about the confidentiality of answers they provide to the census," a rate significantly lower than the rates for households with young children and incomes below $50,000. According to Waleijko and colleagues respondents' cultural demographics also played a role in these differences. Those "who were not English proficient expressed greater concerns than those who were English proficient (42% versus 23%)."

MISSING MINORITIES

The Census Undercount Among Blacks

Since its inception, blacks in America have been undercounted by the decennial census, resulting in a host of cumulative disadvantages for their families, communities and neighborhoods. Indeed, the problem of the black undercount can be traced to the first census, when the black undercount was intentional. When the Constitution was being drafted in 1787, legislators debated over who would be eligible to be counted and how that decision would affect the apportionment of seats in the House of Representatives. It was such government-sanctioned policies that would later set the stage for gerrymandering. Areas with large enslaved communities benefited from the count in terms of federal funding and political power. At the same time, the inclusion of slaves as three-fifths of a person made it plain who did and did not "count" as a whole person. This "differential counting" in our census means that from the beginning, wealthier, white communities have received "more than their fair share" of political representation and economic resources.

An Urban Institute report estimated that the 2020 census could miss counting 1.1 to 1.7 million blacks (Elliot et al. 2019). The undercount among black men has been particularly severe over the decades, a statistical artifact that underscores the dearth of black men in predominantly black neighborhoods. Official census statistics on black men suggests there are even fewer black men in communities of color than actually exist. For 1990, the Commerce Department estimated "a net undercount of about 4 percent for African Americans" as a whole. The undercount was reduced to "2 percent—around 800,000 people—in the 2000 census, but the 2010 showed no significant change to the Black undercount" despite an improvement in the count for the population as a whole.

In addition to an inordinate undercount among black men, approximately 7 percent of young African Americans were uncounted in the 2010 census, a rate roughly twice that for young non-Hispanic White children. Researchers estimated that in 1990, the net undercount for black children was 8 percent, while that for nonblack children was closer to 3 percent. The undercount of black children means that those that disproportionately need programs like the Supplemental Nutrition Assistance Program, and the National School Lunch Program are denied access to them.

The problems enumerating the black population were complicated in 2020 by the onset of the COVID-19 pandemic. Many observers believe irreparable damage was done to the count among communities of color due to the public health crisis. Before the outbreak occurred, civic, advocacy, and nonprofit groups planned multifaceted outreach campaigns to contact communities of color that included social media, telephone calls, and door-to-door canvassing. With on-the-ground contact eliminated by distancing in the wake of COVID-19, organizations faced unprecedented challenges to ensure historically underrepresented groups were counted.

The Census Undercount Among Hispanics and Latinos

Like blacks, Hispanics face major obstacles in being counted in the decennial canvasses, but the reasons vary somewhat from those of other minorities. For Hispanics, the overriding reason for the official undercount is the fear that participating in the census may lead to problems for them or for those with whom they are acquainted. This is especially true if they have an unauthorized status. Texas state Representative César Blanco noted many residents in his Hispanic dominated El-Paso district already mistrust the government, which suppresses census response levels. Technological challenges add another barrier because more than a quarter of El Paso residents do not have broadband Internet subscriptions, which could further limit a community already at risk of an undercount.

A recent report by the National Low Income Housing Coalition estimated that the 2020 census is likely to miss counting 1.2 to 2.2 million Hispanic residents (NLIHC 2019). Early on, local political leaders and activists expressed concern that Hispanic and Latinx participation in the 2020 census would be suppressed due to the citizenship question President Trump wanted included (Straut-Eppsteiner 2019). Indeed, research shows that immigrants feared responding to the census even if they knew the question was excluded (Seesin and Lilley 2020). The possibility of including a citizenship question, coupled with the Trump administration's antagonistic immigration and citizenship policies undermined the likelihood that Hispanics would participate in the census. In Arizona, the get out the count effort was led by "One Arizona," which reached out to a host of marginalized groups, including Asian-Americans, Native Hawaiians and Pacific Islanders, African Americans, Muslim-Americans, Latinx-Americans, undocumented people, and children age five and younger. Although the Trump administration's proposal to require a citizenship question on the census failed, One Arizona leader said many undocumented people were still afraid to complete the census. Many Hispanics fear that essential information, such as their full names and addresses, could be used against them (Eltohamy 2020).

The notion that participating in the US census may have negative repercussions for some groups ignores historic precedence. In what has been an awkward policy for the Bureau to explain, the agency was culpable in assisting the US Government to locate and intern Japanese Americans during World War II. Although it was against the law, census data was used to relocate and incarcerate more than 120,000 men, women, and children of Japanese descent during World War II. Data gathered by the Census Bureau for the 1940 census was used to locate the residences of Japanese-Americans, for purposes of rounding up household members in order that they could be placed in internment camps (DeAngelis 2019; Aratani 2018).

Researchers at UCLA have concluded that a likely Latino undercount in 2020 could cost Los Angeles County over a half billion dollars in federal funding. Torralba (2020) notes the heightened anti-immigrant rhetoric over the past few years has persuaded Latinos to avoid participating in the census. An inaccurate count puts billions of tax dollars at risk for Latino communities that inordinately require funding for health programs, emergency services, infrastructure, and Medicaid. Clearly, the more Latinos are discouraged from participating in the formal count, the more systemic racism becomes entrenched in US society.

Hispanics have not only gone uncounted in previous decennial censuses, they have also been misclassified, leading to a further underestimation of Latino residents in the United States. Chun (2007) found a substantial undercount of Latino residents in the 2000 census. Chun's research notes that nearly half of the Latino population failed to provide information about their national origin, place of birth or ancestry. Rather, they selected the "Other Hispanic or Latino" box to identify themselves. Using the Census Bureau's more detailed 2000 Public Use Microdata Sample, Chun estimates that the US population of Mexican residents in 2000 was almost 7 percent higher than the 2000 census count of 20.9 million. His estimate for Puerto Rican residents is 4 percent higher than the official tally. The difference was much larger among smaller Latino subgroups. For instance, according to the study's estimates, the census should have tallied 999,561 Dominicans instead of 799,768—a discrepancy of 25 percent.

The Census Undercount Among the Foreign-Born

In addition to racial minorities, the decennial census has a tendency to undercount foreign-born residents. Kaneshiro (2013) finds that members of this group who were uncounted in 1990 were most likely to "fit the stereo-typical image of the 'undocumented immigrant.' This includes cohorts from Africa, Central and South America, Mexico, and the Caribbean who entered the country in periods that made them ineligible for amnesty through the Immigration Reform and Control Act (IRCA)." According to Kaneshiro, undocumented persons were likely to feel threatened by the political environment because the recent passage of IRCA created a climate of fear and distrust among undocumented populations. The reform act may have led undocumented persons to feel threatened by the government (and thereby the census) as well as by xenophobic populations who had become more exposed to anti-illegal-immigrant rhetoric.

After analyzing data on population counts, deaths and emigration rates, Kaneshiro estimates that foreign-born residents aged fifteen to forty-four had an 8.76 percent higher undercount in 1990 compared with 2000. Meanwhile, males tended to have a 9.07 percent higher undercount than females in 1990. Kaneshiro finds that arriving in the United States after the IRCA was passed—when unauthorized immigrants were not eligible for amnesty—"has an even stronger effect (9.16%) on relative undercount."

The Census Overcount of Whites

In contrast to undercounting racial minorities, the Census Bureau has his-torically overcounted the number of whites in the United States. In 2010, the census overestimated the white population by between 67,000 and 1.5 mil-lion individuals. In part, the reasons for the overcount are straightforward. In contrast to other groups, white Americans are more likely to have children attending school away from home, where they can be counted both there and at their parents' residence. Another factor is that whites are more likely than other groups to have more than one residence where they may be counted. According to the Urban Institute, even in a low-risk scenario—in which the canvassing process goes relatively smoothly—states with large shares of white residents will see their populations overcounted. Among the states expected to receive a disproportionate amount of political representation and federal funding are Idaho, North and South Dakota, Wyoming, Iowa, and West Virginia.

But other reasons for the disparity in who gets counted are more nefarious and reflect partisan interference in the census enumeration process. Consider the case of Nebraska. The midwestern state has been experiencing a rapid increase in its Hispanic population, doubling from 2000 to 2017. According to the Urban Institute, under a "high-risk" 2020 census canvassing scenario (one in which the state avoided actively preparing for the census), the estimated undercount for Hispanics is 3.6 percent or 8,200 people. Even under a low-risk scenario, the rapidly growing Hispanic population is likely to be undercounted by as many as 4,500 people.

By comparison, a low-risk scenario for the state's white population means overcounting whites in Nebraska by nearly 1 percent, or by 11,300 people. The disparity in the estimated undercount between whites and Hispanics illustrates why official population statistics matter. It highlights why they are of social consequence and why they deserve sustained critical analysis. The following vignette regarding the differential undercount in Nebraska exemplifies how the census process and the official statistics they produce can be politicized by those in powerful positions and undermine the intent of the Constitutionally-mandated Census Bureau.

In 2019, lawmakers in Nebraska established a bipartisan "Complete Count Committee" (recommended by the Census Bureau for every state) before adjourning for the legislative session. The following week, Republican Governor Pete Ricketts vetoed the bill, helping to ensure the census would continue to benefit white, relatively affluent districts over more impoverished communities of color. The implications are profound for those living in the state. According to a study by the University of Nebraska, every person not counted will cost the state $21,000 in federal funds. According to the Center for Public Affairs Research, a 1 percent undercount in the state could result in the loss of $40 million a year in federal funding (cited in Aliaga-Linares 2019). The conflict that unfolded in Nebraska surrounding a get out the count effort speaks powerfully to the way official statistics function to reinforce institutional racism. More generally, it reveals how official population statistics produced by the Census Bureau can be misused, misinterpreted and politicized.

CONCLUSION

The issues surrounding the population figures produced by the decennial census make it transparent why official statistics matter. The misuse, misinterpretation, and politicization of official population data by powerful interests and the mass media make their economic, political, and social implications critically important and deserving of sustained scholarly attention.

More specifically, official population statistics are significant because they are misused when the government allocates political representatives based on them. According to census data collected in 2020, most counties in the United States have declined in population over the past decade. But many grassroots organizers criticize the numbers, believing the data reflects an undercount among communities of color. If this were the case, redistricting could easily overlook neighborhoods where those groups lie and reduce their voting power. Equally important, an undercount would reroute trillions of dollars in federal funding that derive from census data, reducing money for minority communities that are in greatest need of government assistance. Further, official population statistics matter because they are misrepresented as providing an accurate picture of the US population. As we have seen, census figures have consistently created a misleading image of American society, one that is more white than suggested by empirical reality. And they matter because they are politicized when those in powerful positions use the statistics to suppress the presence and influence of minorities and immigrants in society.

As with the other official statistics discussed in this book, the Census Bureau's count of the US population has important social implications. In particular, the numbers play a crucial role in concentrating our attention in particular directions and away from others. Specifically, the figures serve to influence our perception of who does, and who does not count in American society. The Census Bureau's population statistics matter because they cause many consumers of the data to misinterpret the size and growth of different groups in society. Finally, official population statistics matter because they are frequently politicized to create an impression of the United States that lends sympathy for policies that limit the size of the US immigrant population.

Like the other official statistics discussed in this book, the Census Bureau's count of the American population has important economic implications. Population data are used to distribute $1.5 trillion in aid annually, some of which is misallocated due to multiple and sometimes overlapping miscounts. As documented, the census has traditionally overcounted whites in the population. Historically, the use of census data has resulted in communities of color being denied the benefits of much needed social programs, which further serves to reproduce inequality in America. Indeed, the Bureau's "differential undercount" has resulted in an incalculable accumulation of material resources for whites while denying them to minorities. This process has manifestly contributed to systemic racism in the United States. In that official census figures misallocate trillions of dollars in resources (along with political representation), they play a pivotal role in establishing an unequal playing field for marginalized segments of the population. In brief, the Census Bureau's inability to fully enumerate the population has economic consequences in the form of the perpetuation of institutionalized racism and classism in the United States.

Finally, as with the other official statistics surveyed in this book, official population figures have important political implications. This is the case not only in terms of apportioning formal political representation, but also because the numbers and the agencies that produce them are often politicized. A recent case in point involves former Census Bureau Director Steven Dillingham, who resigned abruptly amid charges he had been leaning on Census Bureau personnel to finalize a report on the number of undocumented immigrants in the United States (Montellaro 2021). According to a memo written by the Commerce Department's inspector general, Dillingham attempted to pressure career employees to make a technical data report about "documented and undocumented persons" a "number one priority" to be produced by January 15, 2021 (Wang 2021). Dillingham's efforts were part of a broader campaign orchestrated by President Trump to exclude undocumented immigrants from the population figures used to allocate congressional districts to states (Psaledakis and Shepardson 2021). After his effort to have a citizenship question was discarded by the Supreme Court in June 2019—the Court having characterized Trump's argument for inclusion of a citizenship question as "contrived,"—he then issued an executive order directing the Census Bureau to determine the number of undocumented immigrants through other means. In what must have been regarded as ironic to many, Dillingham opined as he parted the administration, "the world never needed complete and accurate data more than it does now" (Psaledakis and Shepardson 2021). Beyond the explicit effort to exclude immigrants from the numbers used to allocate congressional seats, in states where conservative political administrations were in power, comparatively few resources were devoted to the 2020 census campaign to "get-out-the-count." Conservatives did not see a political advantage in ascertaining a complete count in 2020, so they expended little money and exerted little effort to secure one.

Chapter Five

Why Official Statistics Matter

In this final chapter, the major social, economic, and political themes surrounding official government statistics are summarized with an emphasis on the key ways they have been misused, misinterpreted and politicized. Of central importance here are the contemporary controversies that each of the official statistical series have engendered. Additionally, the last chapter takes a brief look at a final example of a dubious official statistic—the nations' gross domestic product, as calculated by the US Commerce Department. Like many of the other official statistics covered here, a recurring theme regarding GDP numbers concerns what is and what is not counted. The official gross domestic product, the nation's broadest measure of economic activity, omits a significant part of the overall economy by including only formal, above-ground transactions, while ignoring informal exchanges.

Finally, the concluding chapter ends with a brief consideration of an official statistic produced by another nation, Bhutan's Gross National Happiness (GNH) Index. This discussion, like the section in chapter 1 on the early history of the BLS, demonstrates how official government statistics focus our attention in specific ideological directions and away from others that are equally valuable. While official statistics in the United States tend to reflect the primacy of the capitalist economy and the contemporary neoliberal political order, Bhutan uses its GNHI to keep the nation focused on maximizing a collective sense of happiness and well-being.

A CONCLUDING ILLUSTRATION: THE COMMERCE DEPARTMENT'S GROSS DOMESTIC PRODUCT

The history of measuring a nation's gross domestic product can be traced to the seventeenth century. The gross domestic product (GDP) refers to the monetary value of all goods and services sold in a country over the course of a year. The GDP, produced by the Commerce Department, is the broadest indicator of the size of the US economy. It is widely regarded informally as *the* barometer of economic success and overall well-being in American society. As stated by the US Federal Reserve, "Policymakers, government officials, businesses, economists and the public alike rely on GDP and related statistics to help assess the economy's well-being and to make informed decisions" (Thompson 2020). It is also one of the most widely criticized official statistics because of the way it has been misused as a barometer of societal health, misinterpreted as a comprehensive portrait of the economy, and politicized to reflect the values of powerful interests in society.

In the United States, the Gross Domestic Product concept can be traced to the work of Simon Kuznets (1947). Kuznets was a Nobel Prize–winning economist who served at the National Bureau of Economic Research (NBER). In 1937, he presented his ideas on a gross domestic product measure to Congress (Lepenies 2016). Kuznets' idea was "to capture all economic production by individuals, companies, and the government in a single measure, which should rise in good times and fall in bad" (Dickinson 2011). Kuznets' research was warmly received by Congress which was looking for statistical support for the notion that the United States, faced with the threat of a global war, could simultaneously maintain military production goals while also continuing to meet the demand of consumers for basic goods and services.

Kuznets' GDP concept soon became adopted throughout much of the world as the standard by which to measure a nation's economic vitality (Thompson 2020). At the international 1944 "Bretton Woods" meeting that created the International Monetary Fund (IMF), the World Bank, and the General Agreement on Tariffs and Trade (now the WTO), the decision was reached among attendees to institutionalize GDP as the official barometer of the success of the world's economies. The GDP, though widely criticized in academic circles, continues to be embraced by the neoliberal, Bretton Woods-created organizations. This is how the IMF characterizes the GDP today:

> Economists use many abbreviations. One of the most common is GDP, which stands for gross domestic product. It is often cited in newspapers, on the television news, and in reports by governments, central banks, and the business community. It has become widely used as a reference point for the health of national and global economies. When GDP is growing, especially if inflation is not a problem, workers and businesses are generally better off than when it is not. (Callen 2020)

That the GDP continues to be endorsed as the "reference point for the health of national and global economies" reveals the influence powerful interests have in shaping official statistics in contemporary society. The centrality of the GDP also conveys how official statistics channel our perception of social and economic matters to the neglect of alternate, equally legitimate perspectives.

CRITICISMS OF THE GDP

Since its inception, the use of the Gross Domestic Product has exemplified the way official statistics can be misinterpreted and politicized. Despite being warmly welcomed by most mainstream economists and the business establishment, the GDP was also widely criticized in the way it was being misused. Even the architect of the concept expressed discomfort at the way the concept had been used. The problem was not with Kuznets's conceptualization, but rather with "the full throated support and unquestioned acceptance it received" (Hunt 2021).

Prior to his appearance before Congress, Kuznets was quoted as saying "the welfare of a nation can scarcely be inferred from a measurement of national income" (Hunt 2021). Decades later, Kuznets echoed a similar sentiment about the GDP concept when he remarked that "distinctions must be kept in mind between quantity and quality of growth, between its costs and return, and between the short and the long term. Goals for more growth should specify more growth of what and for what" (Croly 1962). Despite his concerns about its widespread adoption, a majority of nation-states in the world today are using the GDP as the standard of social well-being. Many criticisms of the GDP concept as a measure of economic activity and social welfare have been advanced, including prosaic ones. For example, a rapidly growing population may obscure economic trends in society. A large, growing population does not mean a society is improving materially or socially, but relying on GDP will make it appear so as more transactions are completed. In the following discussion, a number of compelling criticisms aimed at the GDP are reviewed.

OVERLOOKING INFORMAL ECONOMIC ACTIVITY

The GDP measure has been critiqued by many observers over the years, including mainstream macroeconomics textbooks. As a consequence, even some leading proponents of the neoliberal economic agenda are seeking to develop a new, "human development" indicator to more accurately assess the state of societal well-being. The narrow approach of including only legal market transactions, while arbitrarily excluding others, is but one of the concerns observers have with the GDP. Critics highlight that while serving informally as a barometer of the overall quality of a nation-state's vitality, the GDP fails to capture essential elements of a high quality of life. To that end, alternatives to the conventional measure of national well-being have been created, and in the case of Bhutan, officially adopted by the state. The point of these final illustrations is to make it plain the book is not designed to be a comprehensive critique of every key statistical series the government produces. The value of unpaid economic undertakings such as barter, volunteering, or household work is excluded from GDP figures.

Undervaluing Housework

The Bureau of Economic Analysis (BEA) points out that a principal omission Kuznets highlighted in his research on national income accounts was "services of housewives and other members of the family." In the years since Kuznets wrote, the hours men contribute to "household production" have risen, while those of women have declined. But it is still true that the exclusion of household production—of men or women—causes a significant understatement in the level of domestic production (BEA 2012). A 2010 study by the BEA found that if the value of household production were included in national income accounts, it would add $3.8 trillion to the US economy and raise officially measured GDP by 26 percent (BEA 2012). A more recent study drawing upon BLS data, estimates it would cost Americans $2.4 trillion to replace the value of the work they performed at home in 2018 (Gioviantti 2020). The effort to determine the value of unpaid economic activities on a national scale has been a focus of social science research for some time. The results remain problematic with varying results, depending on the methodology used to measure informal work (Brown 1974; Glazer-Malbin 1976; Folbre 2012).

Similarly, although difficult to precisely document, the value to the US economy of volunteering activities is significant and increasing. According to the Corporation for National and Community Service (CNCS), over the past fifteen years, Americans have volunteered more than 120 billion hours of their time to volunteer activities, valued at an estimated $2.8 trillion (CNCS 2020). It is likely that this figure is something of an understatement as volunteer work in 2017 was pegged at nearly $25 per hour in official BLS statistics. According to the nonprofit group Independent Sector, the estimated value of more than 70 million Americans performing charitable work amounts to more than $27.00 per hour (Independent Sector 2020). According to the organization, Americans have been increasing the extent of their volunteer activity in recent years. The number of hours Americans volunteered for charitable work grew by 7 percent between 2018 and 2019 alone (Independent Sector 2020).

THE GDP'S OMISSION OF CRIMINAL ACTIVITY

In addition to overlooking and discounting the economic contributions of volunteers and of charitable activity generally, the GDP fails to record a wide range of criminal activities as part of the national income accounts. In 2008, the UN agreed upon revised guidelines for member nations to follow when generating and presenting national economic statisitics. The new system, known as the System of National Accounts 2008, explicitly recommends that illegal market activity be included in a nation's measured economy. Soloveichik (2019) finds the BEA has yet to implement the recommendation because "of challenges inherent in identifying suitable source data and differences in conceptual traditions." Including illegal activity increased nominal GDP figures in the United States by more than 1 percent in 2017. By category, illegal drug transactions added $108 billion to measured nominal GDP in 2017, illegal prostitution boosted nominal GDP by $10 billion, illegal gambling added $4 billion, and theft from businesses increased nominal GDP by $109 billion (Soloveichik 2019). Although some European nations have begun to account for the value of many aspects of the subterranean economy, the United States has been reluctant to include the value of the informal economy, such as prostitution and drug sales in its national income reports (Reuters 2014; Alderman 2014; Gonzalez 2014).

In short, the United States is neglecting the illicit informal economy and the considerable financial value it would add to national income statistics. More importantly, the lack of uniformity in reporting national income across the globe lends itself to a statistical misinterpretation about the relative economic vitality of nations. Charges of politicizing national income statistics directed at countries that included illegal activities such as prostitution and drug transactions have been commonplace (Backman 2014; Mount 2014). Further, some economic activity excluded as a part of the official economy, is legal in some states, where it involves legitimate economic transactions, and contributes to local income account statistics.

The differences in calculating income accounts between nation-states, as well as measurement differences between localities and states within the United States, highlights the lack of uniformity characteristic of many official government statistical series. It also exemplifies the conceptual arbitrariness underlying many official statistics (such as the requirement that workers make an "active search" before being officially labeled unemployed). Such criteria are bureaucratically expedient for the agencies that produce and circulate them, but in the process, they sacrifice accuracy and meaningfulness in explicating the subject under examination.

The Potential Value of Arms Trafficking to GDP

Illegal arms trafficking is one of the larger areas of criminal activity overlooked by official GDP figures. The trade in illicit arms is a global problem, with the United States serving as the leading exporter of light armaments, including small arms and light weapons sales. These highly trafficked weapons are the same ones that are most likely to be exported and used by global organized crime operatives. All together, illicit sales in small arms account for 10–20 percent of all weapons purchases worldwide (Schroder 2005). The illicit trade in weaponry has become an issue both for US law enforcement authorities and for the nations to which illicit armaments are exported. Parsons and Vargas have noted that illegal arms flowing across US borders are posing significant problems for neighboring nations. They point out that "from 2014 to 2016, across 15 countries in North America, Central America, and the Caribbean, 50,133 guns that originated in the United States were recovered as part of criminal investigations . . . during this span, U.S.-sourced guns were used to commit crimes in nearby countries approximately once every 31 minutes" (2018).

The Potential Value of Illicit Drug Sales to GDP

As an illustration of the potential effect of criminal activity on the GDP, consider the growing impact the sale of marijuana would have on the nation's GDP if all transactions were counted. As of early 2021, thirty-six states had legalized medical marijuana. An additional fifteen states, along with Washington, D.C., have legalized the sale of recreational cannabis for consumers over age 21. With legislation varying by state, assessing an accurate statement of the nation's overall GDP becomes problematic. Further, it makes it difficult to make meaningful comparisons between states.

In recent years, the sale of cannabis has increased rapidly in the United States—a 46 percent increase in 2020 over 2019 (Yakowicz 2021). Despite Americans' changing attitudes and behavior, over 70 percent of all cannabis sales remain black market transactions (Jacimovic 2021; Barcott 2020). Even in states where cannabis sales are legal, like California, black market marijuana is still an expanding business, exceeding the value of licit transactions. The arbitrary distinction between formal and informal exchanges leads to a misinterpretation regarding the total amount of economic activity occurring in the state, and the nation as a whole (Romero 2019). Precisely measuring the entire value of cannabis sales in the United States is a difficult undertaking, but it is plain the official GDP numbers are understated as a result of their exclusion.

Estimating the value of illicit drug sales is a complicated undertaking, but worth considering with respect to its potential impact on the nation's official GDP statistics. Ross (2019) notes that despite a decades long War on Drugs, "Americans continue to spend heavily on illicit substances, buying $150 billion worth of cocaine, heroin, marijuana and meth annually." A 2016 study by Arcview market research found that 87 percent of all pot sales in the United States and Canada were black market transactions, worth $46.4 billion in sales, compared with $6.9 billion in legal trade during the same period (Yacowicz 2021). A Rand Corporation Study found that US consumers spent between $121 billion and $146 billion per year on cocaine, heroin, marijuana, and methamphetamine from 2006 to 2016. Drawing upon multiple sources, the Rand report chronicled the number of illicit drug users in the United States and how their expenditures vary on illicit substances.

For 2006, the Rand study found people spent the most on cocaine and the least on marijuana, but that by 2016, this consumption pattern had reversed. Spending on cocaine fell about 60 percent between 2006 and 2013, before stabilizing at a level of $24 billion in 2016. The same study indicated that users laid out an estimated $44 billion on meth in 2016, an increase of $5 billion compared with 2006, while the amount expended on heroin in the same period advanced by $12 billion to more than $43 billion annually (Midgette et al. 2019). Extrapolating these figures to the broader economy makes it transparent that adding illegal drug transactions would significantly boost the official GDP statistics and lead to a more more accurate assessment of individuals' market behavior and the actual size of the economy. As currently constructed, GDP figures artificially underplay the magnitude of the economy and obfuscate much of the productive and consumptive behavior in which people are engaged.

The Potential Value of Human Trafficking and Sex-Related Crime to GDP

There are many informed estimates concerning the economic significance of prostitution and related sex crimes in the US. The bulk of this research is beyond the scope of this study, but consider a 2007 pioneering report by the US Department of Justice. This investigation, which covered criminal enterprises in eight major cities determined that illegal sex-related activity was valued at between $39.9 and $290 million annually (Dank et al. 2014). A more recent comparative examination of the economic worth of prostitution by Havoscope, (a website dedicated to global black market activity) found that including prostitution would add nearly $15 billion to the US GDP (2020).

An alternative possibility about the outcome of including illicit activities on income accounts has been advanced. If black market activities, especially so-called "victimless" crimes were brought above ground and normalized, their contribution to the GDP may be lessened compared to the role they formerly played in the underground economy. For instance, if heroin addicts had access to legal dispensaries to purchase their goods, the price would likely drop as their reliance on illicit street dealers would decrease. This in turn would reduce the overall value of real economic activity and depress GDP statistics.

Even from a cursory examination of these three criminal categories that are unrecognized by US GDP statistics, it is unmistakable that failing to include illegal economic activity has a significant impact on official statistics. With both national economic policies and the careers of politicians at stake, the tendency on the parts of many to massage official economic accounting statistics presents a formidable temptation. The omission of criminal activities and its depressing impact on GDP statistics is but one of many criticisms that have been leveled at these authoritative numbers.

OTHER CRITIQUES OF THE GDP

The Uneven Accounting of Goods and Services

Apart from overlooking criminal endeavors, another central criticism directed at GDP statistics is the uneven job the metric does of assessing different types of economic transactions. Specifically, the existing national accounts system effectively tracks exchanges involving goods. But it is not nearly as reliable at recording the value of transactions involving services. While goods typically have an accompanying manufacturers' recommended sales price that facilitates recording final transactions in GDP statistics, the value of services can only be calculated by assigning an arbitrary quantitative metric someone receiving a service. Although both types of calculation involve an element of subjective evaluation, the measurement of service transactions heavily depends on a wider range of personal tastes.

The GDP's Emphasis on Quantity and Its Neglect of Growing Inequality

The GDP's emphasis on presenting the total value of all economic transactions has been criticized for being short-sighted in accenting quantity rather than quality. This is an incisive example of how official statistics can serve to focus attention in certain directions and away from others. The GDP directs our attention toward the value of material goods, rather than the quality of goods and services. In short, the GDP has frequently been critiqued for implying that bigger is better. Recent experiences such as the Global Financial Crisis and the accompanying Great Recession demonstrate that the increasing size of the financial sector in the international economy has had a detrimental impact on social health and development.

Moreover, the GDP is an aggregate measure, a reflection of what is transpiring in the overall economy. It provides no information about how gains in national income are being distributed. At a time of rapidly increasing economic inequality, news of a growing GDP sends a message that is easily misinterpreted. Despite income gains being inordinately concentrated among the upper part of the distribution, it is commonly maintained that social and economic conditions are improving for everyone when the GDP is advancing. Using the COVID-19 public health crisis to make a critical point about national income accounts, Macekura notes:

Our decades-long obsession with growth has masked an economy that has grown less fair and less capable of providing a good life for people all around the world. Gross domestic product, which measures the monetary value of all goods and services produced within a country, is a poor measure of economic and societal health. During the 2010s, the GDP rose, but without lifting all Americans or even most Americans to a better life. A weak recovery from the 2008–2009 financial crisis lurked below gradual GDP increases. COVID-19 has hammered this message home. (2020)

Finally, the GDP has been roundly condemned because it inflates national accounts by including events that most observers would regard as detrimental to the well-being of society. Pilling states: "GDP is a gross number. It is the sum total of everything we produce over a given period. It includes cars built, Beethoven symphonies played and broadband connections made. But it also counts plastic waste bobbing in the ocean, burglar alarms and petrol consumed while stuck in traffic" (2018).

As a case in point, consider that when the Mississippi River floods each spring, the cost of reconstructing communities from the "natural disaster," including the cost of rebuilding homes, businesses and physical infrastructure in the aftermath are added to official GDP numbers. In 2019, damage from floods approached $4 billion in the United States (Statista 2021). Overall, natural disasters cost the US an unprecedented $306 billion in 2017, all of which became a part of the GDP (CBS News).

Similarly, when the US experiences one of its episodic mass shootings, the cost of caring for victims is treated positively in terms of how the economy, and society, are assessed. But few believe floods, tornadoes, or mass violence are signs of a healthy, sustainable society. Perhaps the best illustration of this criticism can be seen with regard to the natural environment. When examining pollution and GDP statistics, the problem is that pollution-generating economic activities as well as efforts undertaken to repair the environment are included and serve to boost national income. This, however, plainly leads to a misuse of the national income accounts. Through including pollution and other negative externalities, the GDP encourages the perpetuation of the belief that all economic activity is valuable. As former Senator Kennedy once said of the national income accounts data, "GDP measures everything, except that which makes life worthwhile" (cited in Stiglitz 2020).

GDP's Neglect of Time and Efficiency

A significant and often overlooked criticism regarding GDP statistics is that they do not account for how time is used. As an aggregate measure that focuses on the value of all transactions in the above ground economy, the GDP neglects the value of human capital inputs in the production process. Two nations may have very similar overall levels of gross domestic product, but in one, it may take eight hours of labor, on average, and another ten to twelve hours to achieve. For most workers, this difference in work versus leisure time makes a major, palpable impact on their overall well-being, yet the GDP neglects this critical aspect.

That the GDP fails on many fronts as a measure of human development is a widely shared belief, even among many individuals allied with the contemporary neoliberal global economic agenda. For example, at the 2016 World Economic Forum, International Monetary Fund head Christine Lagarde, Nobel-prize winning economist Joseph Stiglitz, and MIT professor Erik Brynjolfsson concluded that, "GDP is a poor way of assessing the health of our economies and we urgently need to find a new measure" (Thoma 2016). Due to the significant number of serious omissions involved in calculating the official GDP, a number of alternatives have been advanced by those seeking a comprehensive measure of national development and well-being. Several indicators have been advanced, including the UN's Index of Sustainable Economic Welfare, and the University of Maryland's Genuine Progress Indicator. Perhaps the most frequently cited example of a nation-state adopting an alternative social welfare measure is Bhutan and its Gross National Happiness Index. The case of Bhutan and its GNH index shows there are viable alternatives to the existing series of official national income statistics generated by most modern nations.

BHUTAN'S GROSS NATIONAL HAPPINESS INDEX

In Bhutan, the notion of "gross national happiness" (GNH) as a development philosophy dates to 1972 when the nation's fourth king, Jigme Singye Wangchuck proclaimed the nation would pursue happiness in its path toward development, rather than measuring progress solely through its gross domestic product. The Gross National Happiness Index is based on four over-arching pillars—good governance, sustainable socioeconomic development, cultural preservation, and environmental conservation. To create the index, the four pillars are divided into nine domain headings, including psychological well-being, health, education, time use, cultural diversity and resilience, good governance, community vitality, and ecological diversity and resilience (OPHI 2020). The nation's Centre for Bhutan Studies revised and released an updated GNH index in 2011 that integrated thirty-three social and economic indicators under the nine domains. The GNH index adapted the "Alkire-Foster" multidimensional measurement for the purpose of creating its index, which resulted in the identification of four groups of people—unhappy, narrowly happy, extensively happy and deeply happy (OPHI 2020). In 2008, the Gross National Happiness Index became institutionalized as the official statistic by which the Bhutanese Constitution evaluated the nation's progress. Three years later, the UN unanimously adopted a General Assembly resolution, introduced by Bhutan with support from sixty-eight member states, calling for a "holistic approach to development" aimed at promoting sustainable happiness and well-being (UN News 2011; Kelly 2012).

While the Gross National Happiness Development Agenda has been embraced by many nation-states as a more inclusive measure of societal well-being, the Bhutanese Government has been accused of using the GNH index to deflect attention from its largely unknown history of mistreating religious minorities (March 2016). According to Human Rights Watch (2008), in the name of one of the four key pillars of the GNH philosophy (cultural preservation) over 100,000 people of Nepalese or Hindu origin were expelled. These religious minorities were exiled because they resisted assimilation into the majority Buddhist culture. Although the nation's GNH Index shows that most of its citizens are narrowly, extensively or deeply happy, it fails to report on those they no longer count—the refugee population who fled to Nepal and India prior to the new Constitutional Government assuming power.

Since Bhutan adopted the Gross National Happiness Index over GDP as its premier indicator of social well-being, the concept has attracted global attention. Empowered by a host of academic and political leaders, a modern-day "political happiness movement" has emerged. In 2011, the United Nations General Assembly passed Resolution 65/309 "Happiness: "Towards a

Holistic Approach to Development." The resolution encouraged member nations to follow Bhutan's example—to measure and make happiness a part of the international development agenda. In 2012, the UN General Assembly passed Resolution 66/281, proclaiming March 20 as the International Day of Happiness. The following year the first international observation of the newly-created Day of Happiness was held. Since that time, advocates in many nation-states have begun to experiment with the idea of integrating happiness into official development statistics.

In the United States, the movement to promote gross national happiness has been assisted by a group known as Gross National Happiness USA (GNHUSA). The stated goal of the organization is to increase personal happiness and collective well-being by changing how the United States measures progress and success. The state of Vermont, due in part to a GNHUSA campaign, declared April 13 (Thomas Jefferson's birthday) "Pursuit of Happiness Day" and became the first state in the nation to pass legislation enabling development of alternative indicators to assist in policymaking.

In 2017, GNHUSA began the process of enlisting all fifty states to work on well-being initiatives. As the example of Bhutan and other alternative measures indicate, the existing GDP is not a measure of social welfare and has never served effectively as one. As noted throughout this manuscript, official government statistics have important implications—economically, politically, socially and ideologically. The example of the GDP is particularly significant in terms of how official statistics can function to frame reality for consumers of the data. The GDP figures generated by the US are exclusively concerned with individuals and activities in the formal economy. From the analytical frame provided by official government statistics, those who toil full time keeping house, or charitably dedicate their time to activities that serve others, or who are engaged in nefarious criminal activities do not count. The truncated misimpression GDP statistics provide overlooks the lived experiences of a significant and growing portion of the population.

CONCLUSION

Helping readers to understand the social significance of official statistics and the ways they can be misused, misinterpreted and politicized has been the fundamental objective of this book. Beginning with the opening chapter's discussion of the criticism leveled at the CDC's COVID-19 fatality data, through the final chapter on the census and the way the decennial undercount of minorities contributes to systemic racism, the emphasis has been on the economic and political implications of the use of official statistics. As documented through the chapters, each of the statistical series examined here have multiple social ramifications.

For example, this text reveals that official unemployment statistics are socially significant and worthy of sustained scholarly attention because they are misused (as a proxy for overall labor market conditions), are misinterpreted (as an accurate gauge of all forms of joblessness in the labor force), and are politicized by actors across the political spectrum seeking to manipulate official data for their own or allied interests. Unemployment statistics also serve to frame a misleading impression about work life in the United States, specifically regarding those who are officially counted as unemployed. Because only official work is recorded, some groups in society such as racial minorities, may appear to be less industrious and work less than official statistics reveal, reinforcing stereotypes that have no basis in empirical reality. Contrary to being idle, it is often minorities and immigrants who are disproportionately engaged in arduous, unpleasant informal jobs, yet their contribution is officially ignored.

In addition, this book has shown that official life expectancy statistics matter and are deserving of thoroughgoing criticism because they have been widely misused as a true indicator of the extent to which modern science has extended the average number of years Americans can expect to live. The CDC's life expectancy statistics matter because they have been misrepresented as an accurate barometer of how much longer Americans are living, complicating the kinds of decisions people make about retirement. Additionally, and more critically, official life expectancy statistics matter because they have been politicized; a process that has already facilitated the ability of conservative advocates of the elimination of Social Security and pensions to delay the full retirement age for Social Security beneficiaries from age sixty-two to sixty-seven.

Further, this book has illustrated the social significance of official crime statistics in the way they have been misused, misinterpreted and politicized. As currently constructed, official crime statistics are routinely misinterpreted as a measure of the totality of crime in society, when the Crime Index that gets publicized is limited to eight violent offense and property-related violations. Though widely cited, the FBI's annual index is far from being a comprehensive indicator of all crime and it is socially significant in the way it draws attention of consumers of the data to street crime rather than toward white-collar crimes such as fraud and embezzlement. As stressed in chapter 3, it is white-collar and corporate crimes and the social costs that emanate from them, especially from the ranks of white-collar crimes, that are most costly to society. Moreover, official crime statistics have often been politicized in order to usher in harsh policies advocated by law enforcement authorities who contend they are necessary to combat particular types of crimes; ones often associated with racial minorities or those from lower class backgrounds. Furthermore, official crime statistics matter because to the extent certain types of crimes go underreported, such as crimes against women and children, there will less public awareness and a reduced likelihood that appropriate steps will be taken to address them. From the NCVS, we know that these types of crimes are underreported and that as a consequence an inadequate amount of law enforcement resources are dedicated to reducing the harm from them, including having fewer police to handle initial domestic disputes in minority communities.

Perhaps more than any of the statistical series examined here, the political and economic implications of official government statistics can be gleaned most clearly seen in the case of the Census Bureau and the population data they collect and circulate. The decennial census data, used to determine both political representation and the public funding of local social programs, have very clear and direct implications that are more obvious than others examined in this manuscript. Less obvious is the immediate connection official census data has to the perpetuation of institutionalized racism in the United States. Since its inception, the census, by undercounting minority groups and denying them their fair share of political representation and access to federal funding has gone a great distance toward instilling systemic racism in US society.

References

Abdelmalek, Mark, Josh Margolin, Aaron Katersky and Eden David. 2020. "Coronavirus Death Toll in US Likely Worse Than Numbers Say." *ABC News*. April 7, 2020.

Abdelmalek, Mark, Chris Francescani and Kaitlyn Folmer. "How Accurate Is the US Coronavirus Death Count? Some Experts Say It's Off By 'Tens of Thousands'" *ABC News*. April 30. https://abcnews.go.com/Health/accurate-us-coronavirus-death-count-experts-off-tens/story?id=70385359

Abramovitz, Moses. 1976. "In Pursuit of Full Employment." Eli Ginzberg (ed.) *Jobs for Americans*. Englewood Cliffs, New Jersey: Prentice Hall.

Adams, Drew. 2020. "Pandemic Creates High Risk of Census Undercount in Communities of Color." *NPQ*. May 18, 2020.

AIT. 2015. "Hiding in Plain Sight: The Spiraling Cost of White-Collar Crime." 2015. Actionable Intelligence Technologies. PR Newswire New York. August 19, 2020.

Alabama News Network. 2017. "45th Anniversary of AP Tuskegee Study Story." Alabama News Network. May 10, 2020.

Aliaga-Linares, Lissette. 2019. "On the Importance of Counting Nebraska Latinos in the 2020 Census." https://www.unomaha.edu/college-of-arts-and-sciences/ollas/research/on-the-importance-of-counting-nebraska-latinos-in-the-2020-census-blog-2019.php

Aljazeera. 2020. "Brazil Restores Coronavirus Data After Controversy, Court Ruling." *Aljazeera*. June 9, 2020.

Almeron, Loi. 2020. "Domestic Violence Cases Escalating Quicker in Time Of COVID-19." Mission Local, March 27, 2020.

American Labor Legislation Review. 1921. "Unemployment Survey—1920–1921." 11: 191–220.

American Labor Legislation Review. 1928. "Need For Unemployment Facts Shown by Issues Raised at Washington." *American Labor Legislation Review* 18: 149–62.

Anand, Shuchi, Maria Montez-Rath, Jialin Han, Julie Bozeman, Russell Kerschmann, Paul Beyer et al. 2020. "Prevalence of SARS-CoV-2 Antibodies in A Large Nationwide Sample of Patients on Dialysis in the USA: A Cross-Sectional Study." *Lancet*. October 24, 2020.

Anderson, Margo and Stephen Feinberg. 2002. "Why is There Still a Controversy About Adjusting the Census for Undercount?" *Political Science and Politics* 35 (1): 83–85.

Anson, Pat. 2018. "CDC Admits Rx Opioid Deaths 'Significantly Inflated." *Pain News Network.* March 21, 2018. https://www.painnewsnetwork.org/stories/2018/3/21/cdc-admits-rx-opioid-deaths-significantly-inflated

Aratani, Lori. 2018. "Secret Use of Census Info Helped Send Japanese Americans to Internment Camps in WWII." *Washington Post.* April 6, 2018.

Arias, Elizabeth. 2016. "Changes in Life Expectancy by Race and Hispanic Origin in the United States, 2013–2014." NCHS Data Brief no. 244 (April).

Aschner, Judy L. Jean L. Raphael and Shale L. Wong. 2019. "The 2020 Census and the Child Undercount: A Threat to Pediatric Research and the Health and Wellbeing of Children." *Pediatric Research* 86. (June): 289–90.

Asia News Monitor. "United States: Rural Areas, Tribal Lands Lag in Getting Census Forms." *Asia News Monitor* (Bangkok), May 26.

Associated Press. 2020. "More People Have Died From COVID-19 In New York City Than Perished On 9/11." *Fortune.* April 7, 2020.

Backman, Melvin. 2014. "Britain, Italy Include Drugs and Sex In GDP." *CNNMoney.* May 30, 2014.

Baltimore Sun Editorial Board. 2019. "The Criminalization of Homelessness." *Baltimore Sun.* December 23, 2019.

Bancroft, Gertrude. 1979. "Some Problems of Concepts and Measurement." in Diane Werneke (ed.) *Counting the Labor Force*, Appendix vol 3: 44–52 Washington: U.S. Government Printing Office.

Baragona, Justin. 2020. "Rush Limbaugh Floats Theory That Coronavirus Deaths Are Being Inflated to Push an Agenda." *Daily Beast.* April 2, 2020.

Barcott, Bruce. 2020. "Americans Will Spend $60 Billion On Illicit Marijuana This Year, Report Says." *Leafly.* September 16, 2020.

Barnett, Cynthia. 2000. "The Measurement of White-Collar Crime Using Uniform Crime Reporting (UCR) Data." US Department of Justice. Federal Bureau of Investigation. Criminal Justice Information Services (CJIS) Division.

Barnett-Ryan, Cindy, Lynn Langton and Michael Planty. 2014. "Nation's Two Crime Measures." US. Department of Justice. (September).

Bartash, Jeffry. 2020. "The Record Number of People Applying for Jobless Benefits Is Even Worse Than It Looks." Marketwatch.com. May 8, 2020.

Basu, Moni. 2012. "U.S. Broadens Archaic Definition of Rape." *CNN.* January 6, 2012.

Bawley, Dan. 1982. *The Subterranean Economy.* New York: McGraw-Hill.

Beasley, Deena. 2020. "U.S. COVID-19 Deaths Likely Higher Than Reported, Study Shows." *Reuters.* July 1, 2020.

Beattie, Ronald H. 1955. "Problems of Criminal Statistics in the United States." *Journal of Criminal Law and Criminology* 46 (2).

Beer, Tommy. 2020. "Actual Number of Coronavirus Deaths Is Likely Far Higher Than Official Tally, Studies Suggest." *Forbes.* July 2, 2020.

Belluz, Julia. 2018. "The New CDC Director Was Once Accused of Research Misconduct." Vox. March 22, 2018.

Berry, Maya and Kai Wiggins. 2018. "FBI Stats on Hate Crimes Are Scary. So Is What'S Missing." CNN Opinion. November 14, 2018.

Best, Joel. 2012. *Damned Lies and Statistics.* MJF Books.

Biderman, Albert D. and Albert J. Reiss, Jr. 1967. "On Exploring the 'Dark Figure' of Crime." *The Annals of the American Academy of Political and Social Science.* 374, Combating Crime (Nov): 1–15.

Blake, Aaron. 2020. "Birx and Fauci Reject Fox News-Promoted Theory That Coronavirus Deaths Are Inflated." *Washington Post.* April 8, 2020.

Blidner, Rachelle. 2020. "Child Abuse Cases Underreported During Pandemic." Newsday. May 13, 2020.

Bonn, Scott. 2017. "Why Elite White-Collar Criminals Are Rarely Punished." *Psychology Today.* April 9, 2017.

Borah, Rohit. 2015. "The Story of Dr. Joseph Moutin and the Foundation of American Public Health." https://medium.com/@rborah/dr-joseph-mountin-public-health-policy-visionary-ccad2bb34a0a

Bosworth, Barry, Gary Burtless and Kan Zhang. 2016. "Later Retirement in Old Age, and the Growing Gap in Longevity Between Rich and Poor." *Economic Studies at Brookings.*

Bradbury-Jones, Caroline and Louise Isham. 2020. "The Pandemic Paradox: The Consequences Of COVID□19 On Domestic Violence." *Journal of Clinical Nursing.* 29 (July) 13–14: 2047–49.

Branswell, Helen. 2018. "CDC Director Brenda Fitzgerald Resigns After Report She Invested in Tobacco Stocks." January 31, 2018. https://www.statnews.com/2018/01/31/brenda-fitzgerald-cdc-resigns/?utm_campaign=rss

Brown, Marie. 1974. "Sweated Labour: A Study of Homework." London: Low Pay Unit 1979.

Brueck, Hilary and Shayanne Gal. 2020. "How the Coronavirus Death Toll Compares to Other Pandemics, Including SARS, HIV, and the Black Death." *Business Insider.* May 22, 2020.

Brusuelas, Joseph. 2020a. "CHART OF THE DAY: What's the Real Unemployment Rate?" Real Economy Blog. October 22, 2020.

Brusuelas, Joseph. 2020b. "CHART OF THE DAY: Estimating Unemployment During the Pandemic." Real Economy Blog. December 4, 2020.

Bureau of Labor Statistics. 1929. *Handbook of Labor Statistics*, 1929 Edition. BLS Bulletin No. 491. Washington: U.S. Government Printing Office.

Burke, Alison S., David Carter, Brian Fedorek, Tiffany Morey, Lore Rutz-Burri, and Shanell Sanchez. 2019. "Street Crime, Corporate Crime, and White-Collar Crime." *Introduction to the American Criminal Justice System.* https://openoregon.pressbooks.pub/ccj230/

Business Week. 1955. "A Closer Count of the Jobless." 148–50. August 6, 1995.

Butterfield, Fox. 2002. "Some Experts Fear Political Influence on Crime Data Agencies." *New York Times.* September 22, 2002.

Byanyima, Winnie. 2016. "Let's Ditch the Economy of the 1% And Replace It with A Human Economy." World Economic Forum. April 20, 2020.

Callen, Tim. 2020. "Gross Domestic Product: An Economy's All." *International Monetary Fund*. February 24, 2020.

Cancryn, Adam and Jennifer Haberkorn. 2018. "Why the CDC Director Had to Resign." *Politico*. January 31, 2018.

Capps, Kriston. 2020. "Where a Census Undercount Will Hurt (or Help) Most." *Bloomberg CityLab*, June 5, 2020.

Cartwright, Alexander. 2017. "GDP Is a Tool of Politics, Not Economics." *Quarterly Journal of Austrian Economics* 20 (1) October 18, 2017.

Case, Anne and Angus Deaton. 2015. "Rising Morbidity and Mortality in Midlife Among White Non-Hispanic Americans in the 21st Century." *Proceedings of the National Academy of Sciences*. 112 (49).

Castillo, Elizabeth. 2020. "Coronavirus and the Census: California Fears an Undercount." CalMatters. San *Jose Inside*. March 17, 2020.

Castillo, Elizabeth. 2020. "How Coronavirus Is Busting California's $187 Million Census Campaign." *Capradio*. May 1, 2020.

Centers for Disease Control and Prevention. 2018. "Malaria: CDC's Origins and Malaria."

Chatterjee, Patralekha 2020. "Is India Missing COVID-19 Deaths?" *Lancet*. September 5, 2020. DOI: https://doi.org/10.1016/S0140-6736(20)31857-2.

Chetty, Raj. 2016. "The Association Between Income and Life Expectancy in the United States, 2001–2014." *Journal of the American Medical Association* 315 (16): 1750–66.

Chodorov, Frank. 1959. "Pernicious Unemployment." *Freeman* 9: 3–7.

Choldin, Harvey. 1994. *Looking for the Last Percent: the Controversy over Census Undercounts*. Rutgers University Press.

Chun, Sung-Chang. 2007. "The 'Other Hispanics'—What are Their National Origins?: Estimating the Latino-Origin Populations in the United States." *Hispanic Journal of Behavioral Sciences*. https://doi.org/10.1177/0739986306297441.

Clague, Ewan. 1961. "How Many Are Really Unemployed'?" Interview in U.S. News and World Report 51: 80–85.

Clague, Ewan. 1968. *The Bureau of Labor Statistics*. New York: Praeger.

Clark, Dartunorro. 2019. "'Money and Power': Fearing an Undercount, States and Cities Pour Millions Into 2020 Census." *NBC News*. October 5, 2019.

Clark, Eleanor. 2018. "Experts: Crime Statistics in Political Ads May Mislead Voters." *WFSU News*. October 2, 2018.

Clinard, Marshall B. 1952. *The Black Market: A Study of White-Collar Crime*. Montclair, N.J.: Patterson Smith.

Clinard, Marshall B. and Peter C. Yeager. 1980. *Corporate Crime*. New York. Free Press.

Cobb, Clifford, Ted Halstead, and Jonathan Rowe. 1995. "If the GDP Is Up, Why Is America Down?" *Atlantic*. October.

Cohen, Mark A. 2013. "Economic Costs of White-Collar Versus Street Crime." *American Association for the Advancement of Science Annual Meeting*. February.

Collinson, Patrick. 2015. "Wealthy Men Living Longer Than Average for First Time." *Guardian*. October 21, 2015.

Condensed Science. 2011. "Life Expectancy in Hunter Gatherers." June 28, 2011.

Council of Economic Advisers. 2018. The Cost of Malicious Cyber Activity to the US Economy. (February). Executive Office of the President of the United States.

Cox, Jeff. 2020. "An Unemployment Rate Of 23%? The Real Jobless Picture Is Coming Together." *CNBC*. April 23, 2020.

Cox, Jeff. 2020. "Weekly Jobless Claims Counts Are Inaccurate and the Unemployed Are Being Underpaid, Watchdog Says." *CNBC*. November 30, 2020.

Croly, Herbert David. 1962. *The New Republic* 147: 29.

Cross, Austin. 2019. "Yes, the Census Bureau Helped Make Japanese American Internment Possible." LAist, June 11, 2019.

Daily Express. 2014. "Life Expectancy for Men Now Higher Than for Women In 100 Areas." https://www.express.co.uk/life-style/health/460166/Life-expectancy-for-men-now-higher-than-for-women-in-100-areas

Daniels, Joseph. 2020. "Have Child Abuse Cases Could Be Underreported Due to Coronavirus Pandemic." *ABC* 10. June 26, 2020

Dank, Meredith, Bilal Khan, P. Mitchell Downey, Cybele Kotonias, Debbie Mayer, Colleen Owens, Laura Pacifici, and Lilly Yu. 2014. "Estimating the Size and Structure of the Underground Commercial Sex Economy in Eight Major US Cities." *Urban Institute Research Report*.

Darcy, Oliver. 2020. "Right-Wing Media Suggests Coronavirus Death Toll Is Inflated, Despite Experts Saying the Opposite." *CNN Business*. April 7, 2020.

DeAngelis, Tori. 2019. "The Legacy of Trauma." *American Psychological Association. Monitor on Psychology*. 50, no. 2 (February).

Decker, Eileen. 2020. "Full Count?: Crime Rate Swings, Cybercrime Misses and Why We Don't Really Know the Score." https://jnslp.com/2020/02/13/full-count-crime-rate-swings-cybercrime-misses-and-why-we-really-know-the-score/

Dickinson, Elizabeth. 2011. "GDP: A Brief History: One Stat to Rule Them All." *Foreign Policy*. January 3, 2011.

Dong, Xiao, Brandon Milholland and Jan Vijg. 2016. "Evidence for A Limit to Human Lifespan." https://www.nature.com/articles/nature19793

Donovan, Sarah A. 2015. "An Overview of the Employment-Population Ratio." Congressional Research Service. May 27, 2015.

Druzin, Heath. 2020. "Gun Rights Group in Idaho Pushes for Looser Firearm Restrictions." *NPR*. March 12, 2020.

Ducharme, Louis Marc, James Tebrake and Zaijan Zhan. 2020. "Keeping Economic Data Flowing During COVID-19." IMF Blog. May 26, 2020.

Dunker, Chris. 2019. "Rickets Vetoes Bills After Session Wraps; No Overrides Possible." *Lincoln Journal Star*. June 4, 2019.

Dutta, Sanchari Sinha. 2020. "250,000 Cases of Child Abuse or Neglect May Have Gone Unreported in U.S. COVID Pandemic." November 12, 2020. https://www.news-medical.net/news/20201112/250000-cases-of-child-abuse-or-neglect-may-have-gone-unreported-in-US-COVID-pandemic.aspx

Dwyer, Colin. 2020. "Brazil Must Be Open with Its Coronavirus Data, Supreme Court Justice Rules." *NPR*. June 9, 2020.

Dwyer, Dialynn. 2020. "Doctors Condemn Conspiracy Theory Pushed by Trump That COVID-19 Deaths Are Over-Counted by Hospitals." https://www.boston.com/news/coronavirus/2020/10/26/doctors-condemn-conspriacy-theory-coronavirus-trump/

Dyer, Owen. 2020. "Covid-19: Russia Admits to Understating Deaths by More Than Two Thirds." BMJ doi.org/10.1136/bmj.m4975

Dzhanova, Yelena. 2019. "Experts are Worried the Census Will Once Again Undercount Kids Younger than 5." *CNBC*. July 29, 2019.

Eads, David. 2018. "Too Many Politicians Misuse and Abuse Crime Data." *New York Times*. August 10, 2018.

Ecarma, Caleb. 2020. "Fox Hosts Now Convinced Coronavirus Death Tolls Are Being Inflated." *Vanity Fair*. April 8, 2020.

Economist. 2014. "Sex, Drugs, and GDP." *Economist*. May 31, 2014.

Ehrenfreund, Max. 2015. "The Stunning—and Expanding—Gap in Life Expectancy Between the Rich and the Poor." *Washington Post*. September 18, 2015. www.washingtonpost.com/news/wonk/wp/2015/09/18/the-government-is-spending-more-to-help-rich-seniors-than-poor-ones/

Elliot, Farley. 2020. "California Has Been 'Significantly' Undercounting its COVID-19 Cases for Weeks." Eater LA. August 5, 2020.

Elliott, Diana, Rob Santos, Steven Martin and Charmaine Runes. 2019. "Assessing Miscounts in the 2020 Census." *Urban Institute*. June.

Ellison, Charles D. 2017. "Underspending in 2020 Census Could Lead to Undercount of Blacks." *Philadelphia Tribune*. April. 11: 1A, 5A.

Eltohamy, Farah. 2020. "Maricopa County Ranks No. 2 For Potential census undercount, report Says." *Cronkite News*. March 25, 2020.

Farrell, Amy and Jessica Reichert. 2017. "Using U.S. Law-Enforcement Data: Promise and Limits in Measuring Human Trafficking." *Journal of Human Trafficking*. 3 (1): 39–60. http://dx.doi.org/10.1080/23322705.2017.128032

Farrell, Amy. 2020. "Gaps in Reporting Human Trafficking Incidents Result in Significant Undercounting." *National Institute of Justice*. August 4, 2020.

Feagin, Joe R. and Clairece Booher Feagin. 1997. *Social Problems: A Critical Power-Conflict Approach*. Prentice-Hall.

Feige, Edgar. 1979. "How Big is the Irregular Economy?" *Challenge*. November–December, 5–13.

Finch, Michael. 2020. "California Must Pivot Away from Its 'High Touch' Census Plans." *Sacramento Bee*. April 17, 2020.

Fleurbaey, Marc. 2009. "Beyond GDP: The Quest for a Measure of Social Welfare." *Journal of Economic Literature* 47 (4): 1029–75.

Folbre, Nancy. 2012. "Valuing Domestic Product." *New York Times*. May 28, 2012.

Follett, Chelsea. 2020. "Americans Have Always Politicized Public Health." *American Conservative*. May 20, 2020.

Fortune. 2020. "17% of Unemployed Workers Aren't Looking for Work—And That's Warping the Official Unemployment Rate." *Fortune*, May 7, 2020. https://fortune.com/2020/05/07/unemployment-numbers-workers-not-looking-jobs-us-economy

Freeman, Erin. 2020. "Child Abuse in County Increasingly Unreported During Pandemic." *Lynnwood Times*, October 14, 2020.

Frelick, Bill. 2008. "Bhutan's Ethnic Cleansing." Human Rights Watch. https://www.hrw.org/news/2008/02/01/bhutans-ethnic-cleansing#

Frosch, Dan and Zusha Elinson. 2021. "Boulder Shooting Prompts New Calls for Gun Control in Colorado." *Wall Street Journal*. March 24, 2021.

Furman, Jason and Wilson Powell III. 2020. "Alternative Unemployment Measures Reveal Deteriorating US Labor Market." *Peterson Institute of International Economics*. December 4, 2020.

Furman, Jason and Wilson Powell. 2021. "Unemployment Continues to Fall but Workers Still Not Returning to the Labor Force." *Peterson Institute for International Economics*. Febrauary 5, 2020.

Gambino, Lauren. 2017. "Tom Price Resigns as Health Secretary Over Private Flights and Trump Criticism." *Guardian*. September 29, 2020.

Garvey, Megan. 2020. "Los Angeles at Risk of Being Undercounted in 2020 Census." *WBUR*. January 7, 2020.

Gawel, Antonia. 2016. "4 Lessons from Bhutan on the Pursuit of Happiness Above GDP." World Economic Forum. May 3, 2016.

Gillespie, Claire. 2020. "This Is How Many People Die from the Flu Each Year, according to the CDC." *Health*. September 24, 2020.

Giovianetti, Erika. 2020. "Women Still Do More Housework Than Men, Contributing $10K+ in Value Annually." MagnifyMoney. March 9, 2020.

Glazer-Malbin, Nona. 1976. "Housework." Signs: Journal of Women in Culture and Society 1 (4): 905–22.

Glickhouse, Rachel. 2019. "Police Don't Do a Good Job Tracking Hate Crimes. A New Report Calls on Congress to Take Action." *ProPublica.* November 13, 2019.

Goldhill, Olivia. 2021. "Undercounting of Covid-19 Deaths Is Greatest in Pro-Trump Areas, Analysis Shows." STAT. January 25, 2021.

Gorman, Mark. 2011. "Life Expectancy: Myth and Reality." November 21, 2011. www.helpage.org/blogs/mark-gorman-25/life-expectancy-myth-and-reality-374/

Governing: The Future of States and Localities. 2021. "Police Employment, Officers Per Capita Rates for U.S. Cities." https://www.governing.com/archive/police-officers-per-capita-rates-employment-for-city-departments.html

Graham-Harrison, Emma, Angela Giuffrida, Helena Smith and Liz Ford. 2020. "Lockdowns Around the World Bring Rise in Domestic Violence." *Guardian*. March 28, 2020.

Greenberg, Jon. 2020. "COVID-19 Skeptics Say There's An Over-Count. Doctors In the Field Say the Opposite." *Politifact*. April 14, 2020.

Greenblatt, Mark and Mark Fahey, Newsy, Bernice Yeung, ProPublica, and Emily Harris. 2018. "FBI Moves to Fix Critical Flaw in Its Crime Reporting System." Reveal from the Center for Investigative Reporting. December 6, 2018.

Gurven, Michael, and Hillard Kaplan. "Longevity Among Hunter-Gathers: A Cross-Cultural Examination." *Population and Development Review* 33 (2): 321–65. June 2007.

Hale, Kori. 2020. "Being Undercounted in the U.S. Census Costs Minority Communities Millions of Dollars." *Forbes*. March 24, 2020.

Hamer, Emily. "'Unnoticed and Unreported': Child Abuse Reporting Down During Pandemic, While Internet Crimes Increase." *Wisconsin State Journal*. https://madison.com/wsj/news/local/crime-and-courts/unnoticed-and-unreported-child-abuse-reporting-down-during-pandemic-while-internet-crimes-increase/articlecb150e15-fbc8-51f7-94e8-b0c2045e2f5.html

Hartman, Michael. 2021. "Cannabis Overview: Legalization." National Conference of State Legislatures. July 6, 2021.

Haseltine, William. 2020. "Lancet Study Suggests U.S. Is Massively Undercounting Covid-19 Cases." *Forbes*. November 4, 2020.

Hauck, Grace. 2019. "Anti-LGBT hate crimes are rising, the FBI says. But it gets Worse." *USA Today*. July 1, 2019.

Havoscope. 2020. "Prostitution Revenue by Country." https://havoscope.com/prostitution-revenue-by-country

Haynie, Devon. 2020. "Coronavirus Quarantines Spark Drop in Crime—For Now." *US News and World Report*. March 30, 2020.

Heller. Jean. 2017. "AP WAS THERE: Black Men Untreated in Tuskegee Syphilis Study." *Associated Press*. May 10, 2017.

Henry, Stuart. 1978. *The Hidden Economy*. Oxford: Martin Robertson.

Hess, Abigail Johnson. 2018. "Here's How Much More Women Could Earn If Household Chores Were Compensated." *CNBC*. April 11, 2018.

Hillygus, D. Sunshine, Norman H. Nie, Kenneth Prewitt and Heili Pals. 2010. *The Hard Count. The Political and Social Challenges of Census Mobilization*. Russell Sage Foundation, New York.

Hindelang, Michael. 1974. "The Uniform Crime Reports Revisited." *Journal of Criminal Justice*. 2 (1): 1–17. https://doi.org/10.1016/0047-2352(74)90114-7

Hoekstra, Rutger. 2019. *Replacing GDP by 2030: Towards a Common Language for the Well-Being and Sustainability Community*. Cambridge, New York: Cambridge University Press.

Holloway, April. 2014. "The Life Expectancy Myth, and Why Many Ancient Humans Lived Long Healthy Lives." April 24, 2014. *Ancient Origins*. https://www.ancient-origins.net/news-evolution-human-origins/life-expectancy-myth-and-why-many-ancient-humans-lived-long-077889

Hoosier, Michael, Branden Stein and Ephraim (Fry) Wernick. 2020. "How COVID-19 Is Affecting the Enforcement of White-Collar Crime." Vinson and Elkins, JD Supra. April 10, 2020.

Horsley, Scott. 2017. "Unemployment Data Are Often Colored by Politics." *NPR* June 3, 2017.

Huff, Darrell. 1954. *How to Lie with Statistics*. W.W. Norton. New York.

Huffington Post. 2014. "U.S. Women Last in Life Expectancy Among Wealthy Countries, Thanks To Growing Inequality." January 27, 2017.

Hunt, Sebastian. 2021. "The Case Against GDP, Made by Its Own Creator." Gross National Happiness USA. March 5, 2021.

Hutson, Matthew. 2020. "The Trouble with Crime Statistics." *New Yorker*. January 9, 2020.

Iacurci, Greg. 2020. "Here's Why the Real Unemployment Rate May Be Higher Than Reported." CNBC. June 5, 2020.

Independent Sector. 2020. "Value of Volunteer Time." https://independentsector.org/value-of-volunteer-time-2021/

Independent Sector. 2020. "Value of Volunteer Time Rose Nearly 7 Percent in 2019." https://independentsector.org/news-post/new-value-of volunteer-time-2019

Ingram, Julia. "Has Child Abuse Surged Under Covid-19? Despite Alarming Stories from ER's, There's No Answer." *NBC News*. https://www.nbcnews.com/health/kids-health/has-child-abuse-surged-under-covid-19-despite-alarming-stories-n1234713

Jacimovic, Darco. 2020. "21 Astounding Cannabis Industry Statistics for 2020." Atheneum Collective. https://atheneumcollective.com/21-astounding-cannabis-industry-statistics-for-2020

Jacobsen, Linda A., Mark Mather and Liselle York. 2020. "Why Are So Many Young Children Undercounted in the US Census?" Population Reference Bureau. https://scorecard.prb.org/why-are-so-many-young-children-undercounted-in-the-u-s-census/

James, Nathan and Logan Rishard Council. 2008. "How Crime in the United States Is Measured." Congressional Research Service. January 3, 2008.

Jarosz, Beth. 2018. "Citizenship Question Risks a 2020 Census Undercount in Every State, Especially Among Children." Population Reference Bureau. October 5, 2018.

Johnson, Kim and Dallas Williams. 2019. "A 2020 Census Undercount Could Cost Texas $300 Million, More Seats in Congress." High Plains Public Radio. January 7, 2019.

Johnson, Marty. 2020. "More New Yorkers Have Died from Coronavirus This Week Than On 9/11." *Hill*. April 10, 2020.

Kadlec, Dan. 2014. "Americans Are Living Longer Than Ever. And That My Kill Your Pension." www.huffingtonpost.com/2014/01/26/us-women-last-in-life-expectancy_4662083.html

Kanazawa, Satoshi. 2008. "Common Misconceptions About Science II: Life Expectancy." Psychology Today on-line. https://www.psychologytoday.com/us/blog/the-scientific-fundamentalist/200811/common-misconceptions-about-science-ii-life-expectancy

Kaneshiro, Matheu. 2013. "Missing Minorities?: The Phases of IRCA Legislation and Relative Net Undercounts of the 1990 vis-à-vis 2000 Decennial Census for Foreign-born Cohorts." *Demography* 50: 1897–919.

Karlin-Smith, Sarah and Brianna Ehley. 2018. "Trump's Top Health Official Traded Tobacco Stock While Leading Anti-Smoking Efforts." https://www.politico.com/story/2018/01/30/cdc-director-tobacco-stocks-after-appointment-316245

Kelly, Annie. 2012. "Gross National Happiness in Bhutan: The Big Idea from A Tiny State That Could Change the World." *Guardian*. December 1, 2012.

Khan, Saher. 2020. "How the Coronavirus Pandemic Has Affected the 2020 Census." *PBS Newshour*. April 1, 2020.

Kimble, Cameron. 2018. "Sexual Assault Remains Dramatically Underreported." Brennan Center for Justice. October 4, 2018.

Kochhar, Rakesh. 2020. "Unemployment Rate Is Higher Than Officially Recorded, More So for Women and Certain Other Groups." Pew Research Center. June 30, 2020.

Kodjak, Alison. 2016. "CDC Finds Life Expectancy for White Women Has Declined." NPR, heard on "All Things Considered." April 20, 2016. https://www.npr.org/2016/04/20/475015253/cdc-finds-life-expectancy-for-white-women-has-declined

Kozlowski, Hanna, and Kopf, Dan. 2020. "Cornoavirus Could Exacerbate the US Census' Undercount of People of Color." *Quartz*. April 22, 2020.

Krause, R. 2006. "The Swine Flu Episode and the Fog of Epidemics." *Emerging Infectious Diseases* 12 (1): 40–43.

Kreston, Rebecca. 2013. "The Public Health Legacy of the 1976 Swine Flu Outbreak." *Discover Magazine*. September 30, 2013.

Krupp, Daniel. 2012. "Marital, Reproductive, And Educational Behaviors Covary with Life Expectancy." *Archives of Sexual Behavior* 41 (6): 1409–14.

Kurtz, Annalyn, Tal Yellin and Byron Manley. 2020. "14.7% Unemployment Is Tragic, And It Doesn't Even Include Everyone Who's Out of Work." *CNN Business*. May 8, 2020.

Kuznets, Simon. 1934. "National Income, 1929–1932." National Bureau of Economic Research. June 1934.

Kuznets, Simon. 1947. "Measurement of Economic Growth." *Journal of Economic History*. https://www.cambridge.org/core/journals/journal-of-economic-history/article/measurement-of-economic-growth/82B15EDDCD0E62F52E4FFE6E0606C1AB82

Kuznets, Simon. 1955. "Economic Growth and Income Inequality." *American Economic Review* 45 (1): 1–28.

Laden, Greg. 2011. "Falsehood: If This Was the Stone Age, I'd Be Dead by Now." https://www.geekwrapped.com/myth-infographic

Lambert, Lance. 2020. "The 'Real' Jobless Rate Is Much Worse Than the Official Numbers Show." *Fortune*. October 14, 2020.

Lamm, Stephanie. 2019. "With Billions in Federal Funds Riding on the 2020 Census, Dallas Wants to Make Sure Every Resident Counts." *Dallas Morning News*. April 7, 2019.

Landefeld, J. Steven. 2020, "Covid-19's Impact on The Economy: Measuring GDP During a Pandemic." IZA World of Labor. May 29, 2020.

Lauitsen, Janet L. and Daniel L. Cork. 2017. "Expanding Our Understanding of Crime: The National Academies Report on the Future of Crime Statistics and Measurement." *Criminology and Public Policy*. https://doi.org/10.1111/1745-9133.12332

Lebergott, Stanley. 1942. "Measuring Unemployment." *Quarterly Journal of Economics*, (November): 19–30.

Leite, Julia, Fernando Travaglini, and Simone Preissler Iglesias. 2020. "Brazil's Covid Data Blackout Is a Tragedy, Ex-Health Chief Says." Bloomberg. June 6, 2020.

Leonard, Sara. 2020. "How to Calculate the Value of Volunteer Time." Nonprofit Leadership Center. October 5, 2020.

Lepenies, Philipp. 2016. *The Power of a Single Number: A Political History of GDP*. Columbia University Press. New York.

Letts, Stephen. 2018. "The GDP Myth: The Planet's Measure for Economic Growth Is Deeply Flawed and Outdated." *ABC News*. June 2, 2018.

Levine, Sam. 2020. "Coronavirus Upends US Census as Bureau Looks to Save Official Count." *Guardian*. March 30, 2020.

Lewis, Bobby. 2020. "Right-Wing Media Suggested COVID-19 Fatalities Were Overcounted. New Data Suggests a Significant Undercount." Media Matters for America. April 27, 2020.

Lianne, Ciara. 2020. "Group of Former CDC Heads Say No President Has Ever Politicized the Leading U.S. Health Agency the Way Trump Has." July 18, 2020. https://www.marketwatch.com/story/group-of-former-cdc-heads-say-no-president-has-ever-politicized-the-leading-us-health-agency-as-trump-has-2020-07-14

Lohr, Sharon. 2019. "Measuring Crime: Behind the Statistics." CRC Press.

Lohr, Sharon. 2019. Edith Abbott and the Origins of Chicago's Crime Statistics Significance, Volume 16: 2. https://doi.org/10.1111/j.1740-9713.2019.01247.x

Long, Heather. 2021. "Are we Undercounting the Unemployed?" *Washington Post*. (February) 2021: 23.

Lopez, Monxo. 2020. "Beyond the Numbers: The 2020 Census and the Covid-19 Pandemic." Museum City of New York. April 9, 2020.

Lujan, Carol Chiago. 2014. "American Indians and Alaska Natives Count: The US Census Bureau's Efforts to Enumerate the Native Population." *American Indian Quarterly* 38, No. 3 (Summer): 319–41.

Lurie, Peter. 2018. "CSPI Urges Administration Not to Appoint Dr. Robert Redfield, with History of Scientific Misconduct, as CDC Director." March 21. Center for Science in the Public Interest.

Macekura, Stephen. 2020. "As Covid-19 Has Exposed, Our Obsession with Economic Growth Is Harming People." *Washington Post*. November 17, 2020.

Macias, Amanda and Nate Rattner. 2020. "Global Arms Trade Is a Nearly 200 Billion Business and the US Drives Nearly 80% Of It." *CNBC*. February 4, 2020.

Maier, Shana L., Suzanne Mannes, and Emily L. Koppenhofer. 2017. "The Implications of Marijuana Decriminalization and Legalization on Crime in the United States." *Contemporary Drug Problems*. May 8, 2017.

March, Maximillian. 2016. "Bhutan's Dark Secret: The Lhotshampa Expulsion." *Diplomat*. September 21, 2016.

Marcuss, R.D. and R.E. Kane. 2007. "U.S. National Income and Product Statistics: Born of the Great Depression and World War II." Bureau of Economic Analysis: Survey of Current Business: 32–46.

Marinetto, Mike. 2014. "Now GDP Data Reflects the Truth: Drugs and Sex Work Boost the Economy." *Conversation*. June 2, 2014.

Marzulli, John. 2013. "Global Cyber, ATM Heist Nets Thieves $45 From 26 Countries." *New York Daily News*. May 10, 2013.

Mattera, Philip. 1985. *Off the Books: The Rise of the Underground Economy*. St. Martin's Press.

McCrary, S. Van. 1998. "Ethical and Public Health Implications of Under-counting Violent Crime." Health Law & Policy Institute. November 25, 1998.

McKay, Betsy. 2016. "Life Expectancy for White Americans Declines." *Wall Street Journal*, April 20, 2016. https://www.wsj.com/articles/life-expectancy-for-white-americans-declines-1461124861

McKinlay, John B. and Sonja M. McKinlay. 1977. "The Questionable Contribution of Medical Measures to the Decline of Mortality in the United States in the 20th Century." *Milbank Memorial Fund Quarterly: Health and Society* 55 (3): 405–28.

Meeker, Royal. 1930. "The Dependability and Meaning of Unemployment and Employment Statistics in the United States." *Harvard Business Review* 8: 385–400.

Melamed, Samantha and Mike Newell. 2020. "With Courts Closed by Pandemic, Philly Police Stop Low-Level Arrests to Manage Jail Crowding." *Philadelphia Inquirer*.

Miami Herald Editorial Board. 2019. "Be Counted in Census 2020, Even Though Florida Leaders Don't Want (Some of You) To." *Miami Herald*. December 29, 2019.

Midgette, Gregory, Steven Davenport, Jonathan P. Caulkins and Beau Kilmer. 2019. *What America's Users Spend on Illegal Drugs, 2006–2016*. The Rand Corporation. Santa Monica, California.

Milman, Oliver. 2020. "Fauci Dismisses 'Conspiracy Theory' Of Overstated US Covid-19 Death Toll." *Guardian*. April 9, 2020.

Minkel, J.R. 2007. "Confirmed: The U.S. Census Bureau Gave Up Names of Japanese-Americans in WW II." *Scientific American*. March 30, 2007.

Montellaro, Zach. 2021. "Census Bureau Director Stepping Down After Outcry Over Immigrant Count." *Politico*. January 18, 2021.

Moses, Stanley. 1975. "Labor Supply Concepts: The Political Economy of Conceptual Change." *Annals of the American Academy of Political and Social Science* 418: 26–44.

Mount, Ian. 2014. "Spain Gets a Questionable GDP Boost, Thanks To Drugs and Prostitution." Fortune. October 8, 2014.

Mule, Thomas. 2012. Memorandum: DSSD 2010 CENSUS COVERAGE MEASUREMENT MEMORANDUM SERIES #2010-G-01. May 22, 2012.

Munnell, Alicia H. 2020. "Are we Undercounting or Overcounting COVID-19 Deaths?" September 30, 2020.

Murphey, David, Dana Thomson, Lina Guzman and Claire Kelley. 2019. "A Census Undercount Likely Cost Detroit $1.3 Million for Childhood Lead Prevention." Talk Poverty. August 14, 2019.

Murphey, David, Dana Thomson, Lina Guzman and Claire Kelley. 2019. "Undercounting Hispanics in the 2020 Census Will Result in A Loss in Federal Funding to Many States for Child and Family Assistance Programs." Child Trends. August 14, 2019.

Murray, Christopher JL. 2011. "Why is Japanese Life Expectancy So High?" *Lancet*. August 30, 2011.

Myers Jr., Samuel L. 1980. "Why are Crimes Underreported? What Is the Crime Rate? Does It 'Really' Matter?" *Social Science Quarterly* 61, no. 1 (June).

Nakamura, David. 2021. "Attacks on Asian Americans During Pandemic Renew Criticism That U.S. Undercounts Hate Crimes." *Washington Post*. February 22, 2021.

National Academies of Sciences, Engineering and Medicine. 2015. *The Growing Gap in Life Expectancy by Income: Implications for Federal Programs and Policy Responses*. National Academies Press.

National Academies of Sciences, Engineering, and Medicine. 2016. *Modernizing Crime Statistics—Report 1: Defining and Classifying Crime*. Washington, DC: National Academies Press. https://www.nap.edu/read/23492/chapter/2

National Law Center on Homelessness and Poverty. 2019. "Housing not Handcuffs: Ending the Criminalization of Homelessness in U.S. Cities." December.

National Low Income Housing Coalition. 2019. "Report Predicts 2020 Census Could Undercount Millions of Blacks and Hispanics." NLIHC. October 18, 2019.

National Low Income Housing Coalition. 2019. "Report Finds Continuing Rise in Criminalization of Homelessness." NLIHC. December 16, 2019.

NBC NewYork. 2020. "NYC launches $40 'Complete Count Campaign' Ahead of 2020 Census." NBC NewYork. January 14, 2020.

Nearing, Scott 1909 "The Extent of Unemployment in the United States." *Journal of the American Statistical Association* 11: 525–42.

Neel, Joe. 2018. "CDC Director Resigns Because of 'Complex' Financial Entanglements." *NPR*. January 31, 2018.

New Republic. 1922. "Normalcy in Unemployment." October 11, 1922. https://www.unz.com/print/NewRepublic-1922oct11/

New York Times. 1930. "Disputes Hoover on Employment." January 23, 1930. https://www.nytimes.com/1930/01/23/archives/disputes-hoover-on-employment-frances-perkins-says-figures-in-this.html

New York Times. 1930. "Employment Turns Upward, Hoover Reports." January 22, 1930. https://www.nytimes.com/1930/01/22/archives/employment-turns-upward-hoover-reports-changes-for-first-time-since.html

Newkirk, Vann. W. 2018. "Is the CDC Losing Control?" *Atlantic*. February 3, 2018.

Nunez, Gabriella. 2020. "Crime Likely Being Underreported During COVID-19 Pandemic, Seminole Sheriff Says." https://www.clickorlando.com/news/local/2020/08/19/crime-likely-being-underreported-during-covid-19-pandemic-seminole-sheriff-says

Nutting, Rex. 2015. "Opinion: The Misery Index Hit A 59-Year Low, So Why Are We Still Miserable?" Marketwatch. October 16, 2015.

O'Connor, Tom. 2020. "Coronavirus has Killed More in the US Than the War in Afghanistan, Death Toll Soon to Pass 9/11." *Newsweek*. March 29, 2020.

O'Hare, William P. 2014. *The Net Undercount of Children in the 2010 U.S. Decennial Census*. In: Hoque M., B. Potter L. (eds) *Emerging Techniques in Applied Demography*. Applied Demography Series, volume 4. Springer, Dordrecht.

O'Hare, William P. 2015. The Undercount of Young Children in the Decennial Census. Spring Briefs in Population Studies. https://www.springer.com/gp/book/9783319189161

O'Hare, William P. 2019. "Differential Undercounts in the U.S. Census. Who is Missed?" *Springer Briefs in Population Studies*. Open Access Book.

O'Hare, William P. 2019. "Potential Explanations for Why People Are Missed in the U.S. Census." *Springer Briefs in Population Studies*: 123–38.

Olejarz, JM. 2016. "Understanding White-Collar Crime." *Harvard Business Review*. November: 110–11.

Olshansky. S. Jay. March 17, 2005. "A Potential Decline in Life Expectancy in the United States in the 21st Century." *New England Journal of Medicine* 352: 1138–45.

Olshansky, S. Jay. 2016. "Aging: Measuring Our Narrow Strip of Life." Nature. October 5, 2016.

Ordway, Denise-Marie. 2019. "2020 Census: How Undercounts and Over-Counts Can Hurt US Communities." *Journalist's Resource*. A Department of Agriculture Publication.

Oxford Poverty and Human Development Initiative. 2020. "Bhutan's Gross National Happiness Index."

Parsons, Chelsea and Eugenio Weigend Vargas. 2018. "Beyond Our Borders." Center for American Progress.

Pew Research Center. 2020. "The #COVID19 Outbreak Has Presented a Unique Set of Measurement Challenges for Statistical Agencies." Pew Research Center. June 30, 2020.

Pilling, David. 2018. "5 Ways GDP Gets It Totally Wrong as A Measure of Our Success." World Economic Forum. January 17, 2018.

Pohl, Jason, Dale Kasler and Philip Reese. 2021. "California OSHA Undercounts Workers' COVID Illnesses, Deaths." *Sacramento Bee*. February 2, 2021.

Poliquin, Christopher. 2021. "Gun Control Fails Quickly in Congress After Each Mass Shooting, But States Often Act – Including to Loosen Gun Laws." *Conversation*. March 25, 2021.

Psaledakis, Daphne and David Shepardson. 2021. "US Census Bureau Director Resigns Ahead of Schedule." Reuters. January 18, 2021. https://www.reuters.com/article/us-usa-census/u-s-census-bureau-director-resigns-ahead-of-schedule-idUSKBN29N1WG

PublicHealthWatch. 2015. "New Report Shows Widening Life Expectancy Gap Between Rich and Poor." www.nejm.org/doi/full/10.1056/NEJMsr043743\.

Puja Seth, Rose A. Rudd, Rita K. Noonan, Tamara M. Haegerich, "Quantifying the Epidemic of Prescription Opioid Overdose Deaths." *American Journal of Public Health* 108, no. 4 (April 1, 2018): 500–2. https://doi.org/10.2105/AJPH.2017.304265

Radford, Benjamin. 2009. "Human Lifespans Nearly Constant For 2,000 Years." August 21, 2009. https://www.livescience.com/10569-human-lifespans-constant-2-000-years.html

Rainey, Rebecca. 2020. "Record-Breaking Unemployment Claims May Be Vast Undercount." *Politico*. March 26, 2020.

Ramsay, A.M. and R.T. Edmond. 1967. Infectious Diseases. Heinemann. Onlinelibr ary.wiley.com

Raskin, A.H. 1979. *Views On Employment Statistics from the Press, Business, Labor and Congress*. Counting the Labor Force Appendix 7: 378–411. Washington: U.S. Government Printing Office.

Reamer, Andrew. 2018. "Counting for Dollars 2020: The Role of the Decennial Census in the Geographic Distribution of Federal Funds. Report #2: Estimating fiscal costs of an undercount to States." George Washington University Institute of Public Policy. March 19, 2018.

Reamer, Andrew. 2020. "Counting for Dollars 2020: The Role of the Decennial Census in the Geographic Distribution of Federal Funds." George Washington University Institute of Public Policy. April 29, 2020.

Regoli, Natalie. 2019. "17 Key Pros and Cons of NCVS (National Crime Victimization Survey)" Connect US. November 20, 2019.

Reuters Staff. 2014. "It's Official: Drugs, Prostitution Boost Dutch Economy." Reuters. June 25, 2014.

Reuters Staff. 2020. "U.S. Jobless Rate Likely Much Higher Than 14.7%, Labor Department Says." Reuters. May 8, 2020.

Robison, Sophia M. 1966. "A Critical View of the Uniform Crime Reports." *Michigan Law Review* 64, no. 6 (April 1966): 1031–54 https://doi.org/10.2307/1286877.

Romero, Dennis. 2019. "California's Cannabis Black Market Has Eclipsed Its Legal One." *NBC News*. September 20, 2019.

Rosenthal, Miriam D. 2000. "Striving For Perfection: A Brief History of Advances for Perfection: A Brief History of Advances and Undercounts in the US Census." *Government Information Quarterly* 17 (2): 93–234.

Ross, Sean. 2019. "The Economics of Illicit Drug Trafficking." Investopedia. November 14, 2019.

Runes, Charmaine. 2019. "Following a Long History, the 2020 Census Risks Undercounting the Black Population." Urban Wire: Race and Ethnicity, Blog of the Urban Institute. February 26, 2019.

Sabbadini, Linda Laura. 2011. "The Development of Official Social Statistics in Italy with a Life Quality Approach." *Social Indicators Research* 102 (1): 39–46.

Salamon, Lester M., S. Wojciech, and Megan A. Haddock. 2011. "Measuring the Economic Value of Volunteer Work Globally: Concepts, Estimates and a Roadmap to the Future." Annals of Public and Cooperative Economics 82 (3): 217–52. September.

Sarma, Bidish and Jessica Brand. 2018. "The Criminalization of Homeless: Explained." Appeal. June 29, 2018.

Savage, Charlie. 2012. "U.S. to Expand Its Definition of Rape in Statistics." *New York Times*. January 6, 2012.

Sawyer, Wendy. 2020. "How to Find and Interpret Crime Data During the Coronavirus Pandemic: 5 Tips." Prison Policy Initiative. April 24, 2020.

Schneider, Mike. 2020. "Sesame Street Wants to Get Young Children Counted in the Census." Vegas PBS. March 9, 2020.

Schneider, Mike. 2020. "Worries About 2020 Census' Accuracy Grow with Cut Schedule." Associated Press. August 5, 2020.

Schultz, Kai. 2017. "In Bhutan, Happiness Index as Gauge for Social Ills." *New York Times* January 17, 2017. https://www.nytimes.com/2017/01/17/world/asia/bhutan-gross-national-happiness-indicator-.html

Schulz, Myron and William Schaffner. 2015. "Dr. Alexander Langmuir." *Emerging Infectious Diseases* 21, no. 9 (September): 1635–37.

Schweers, Jeffrey. 2020. "Agents Raid Home of Fired Florida Data Scientist Who Built COVID-19 Dashboard." *Tallahassee Democrat*. December 7, 2020.

Schwencke, Ken. 2017. "Why America Fails at Gathering Hate Crime Statistics." *ProPublica*. December 4, 2017.

ScienceDaily. 2009. "'Did the Great Depression have a Silver Lining'? Life Expectancy Increased By 6.2 Years." September 29, 2009. https://epo.wikitrans.net/Life_expectancy

Scott, Roxanne. 2020. "Atlanta Pushes for Accurate Count of Young Children On 2020 Census." *PBS* (WABE Radio, Atlanta). March 3, 2020.

Seesin, Carmen and Sandra Lilley. 2020. "Most Latinos Worry the Trump Administration Could Use Census Against Them, New Study Finds." *NBC News*. February 10, 2020.

Seeskin, Zachary H., and Bruce D. Spencer. 2015. "Effects of Census Accuracy on Apportionment of Congress and Allocations of Federal Funds." Working paper from the Institute for Policy Research at Northwestern University.

Sharpe, Robert. 1988. *The Cruel Deception*. Thorsons Publishing Group. Wellingborough, U.K.

Silverstein, Jason. 2020. "President Trump Accuses New York City of Inflating Its Coronavirus Death Toll." *CBS News*. April 16, 2020.

Simon, Carl P. and Ann With. 1982. *Beating the System: The Underground Economy*. Boston: Auburn House.

Simpson, Jack, Sophie Mitra, Elaine Unterhalter, and Jay Drydyk. 2020. "Stop Obsessing Over GDP When Talking About the Pandemic Recovery." *Business Insider*. August 9, 2020.

Singer, James W. 1979. "Big Money at Stake in Redefining Unemployment." *National Journal* 11: 9–13.

Sirota, David. 2015. "US Prosecution of White-Collar Crime Hits 20-Year Low." *International Business Times*. August 4, 2015. https://www.ibtimes.com/us-prosecution-white-collar-crime-hits-20-year-low-report-2037160

Skogan, Wesley. 1974. "The Validity of Official Crime Statistics." *Social Science Quarterly* 55 (1): 25–38.

Skogan, Wesley G. 1977. "Dimensions of the Dark Figure of Unreported Crime." *Crime and Delinquency*. January 1. https://doi.org/10.1177/001112877702300104

Slattery, Denis. 2019. "New York Will Dedicated $40 Million To 2020 Census Outreach in State Budget." *New York Daily News*. March 27, 2019.

Smith, Matt and Lance Williams. 2020. "Fear of a Census Undercount." Reveal. February 1, 2020.

Snowdon, John and Namkee G. Choi. 2020. "Undercounting of Suicides: Where Suicide Data Lie Hidden." *Global Public Health* 15 (12): 1894–901.

Soloveichik, Rachel. 2019. "Including Illegal Activity in the U.S. National Economic Accounts." Bureau of Economic Affairs. July.

Southern Poverty Law Center. 2006. "Report: FBI Hate Crime Statistics Vastly Understate Problem." January 31, 2006.

Sperling, Gene B. 2017. "Government Economists Are Going to Produce Statistics Trump Doesn't Like." *Atlantic*. February 2, 2017.

Staropoli, Nick. 2015. "A Reader Asks, Is Life Expectancy in America Declining?" ASCH.org.

Steinberg, Jake. 2020. "With Census Field Operations Wrapped, Concerns Over Count Persist." Arizona Public Media (NPR). October 22, 2020.

Stewart, Charles and A.J. Jaffee. 1979. "Manpower Resources and Utilization." Diane Werneke (ed.) *Counting the Labor Force Appendix*. 3: 5–12.

Stewart, Emily. 2015. "White-Collar Crime Costs Between $300 To $600 Billion A Year." ValueWalk. July 9, 2015.

Stickle, Ben and Marcus Felson. 2020. "Crime Rates in a Pandemic: The Largest Criminological Experiment in History." *American Journal of Criminal Justice* 45: 525–36.

Stiglitz, Joseph E., Jean-Paul Fitoussi and Martine Durand (Eds.). 2019. *For Good Measure: An Agenda for Moving Beyond GDP*. New York: New Press.

Stiglitz, Joseph E. 2020. "GDP Is the Wrong Tool for Measuring What Matters." *Scientific American*, August 1, 2020.

Stokes, Andrew C., Dielle J. Lundberg, Irma T. Elo,,2 Katherine Hempstead,,3 Jacob Bor, and Samuel H. Preston. 2021. "Assessing the Impact of the Covid-19 Pandemic on US Mortality: A County-Level Analysis." PubMed Central. January 12, 2021.

Strane, Douglas and Heather M. Griffis. 2018. "Inaccuracies in the 2020 Census Enumeration Could Create a Misalignment Between States Needs." *American Journal of Public Health*.

Strane, Douglas. 2020. "COVID-19 and Systemic Racism Threaten the 2020 Census." Children's Hospital of Philadelphia. PolicyLab. October 14, 2020.

Straut-Eppsteiner, Holly. 2019. "Research Shows a Citizenship Question Would Suppress Participation among Latinxs and Immigrants in the 2020 Census, Undermining Its Reliability." National Immigration Law Center. April 22, 2019.

Strmic-Pawl, Hephzibah, Brandon A. Jackson and Steve Garner. 2017. "Race Counts: Racial and Ethnic Data on the U.S. Census and the Implications for Tracking Inequality." *Sociology of Race and Ethnicity*. December 4. https://doi.org/10.1177/2332649217742869

Sullivan, Teresa A. 2020. "Census 2020: Understanding the Issues." *Springer Nature*. Switzerland AG.

Sullivan, Teresa A. 2020. "Coming to our Census: How Social Statistics Underpin our Democracy (and Republic)." *Harvard Data Science Review* Issue 2. (Winter).

Sutherland, Edwin. 1940. "White Collar Criminality." *American Sociological Review* 40 (5): 1.

Sutherland, Edwin H. 1949. *White Collar Crime*. New York: Holt.

Swan, Jonathan and Sam Baker. 2020. "Trump and Some Top Aides Question Accuracy of Virus Death Toll." *Axios*. May 6, 2020.

Tavernise, Sabrina. 2016. "Disparity in Life Spans of the Rich and the Poor Is Growing." *New York Times*. February 12, 2016. https://www.nytimes.com/2016/02/13/health/disparity-in-life-spans-of-the-rich-and-the-poor-is-growing.html

Taylor, Marisa. 2018. "Research Misconduct Allegations Shadow New CDC Head." *Kaiser Health News*. March 21, 2018.

Taylor, Richard. 1979. *Medicine Out of Control*. Sun Books.

TCR Staff. 2017. "Measuring the 'Dark Figure' of Crime." *The Crime Report.*

Terkel, Amanda. 2012. "Eric Holder Expands FBI's Narrow, Outdated Definition of Rape." *Huffington Post*. January 6, 2012.

Thompson, Vanessa. 2020. "The Tragic Simplisticity of GDP." *Berkeley Economic Review*. April 23, 2020.

Time. 1972. "Medicine: A Matter of Morality." August 7, 1972.

Tippett, Rebecca. 2020. "Census Undercounts Are Normal, But Demographers Worry This Year Could Be Worse." *GCN*. April 3, 2020.

Torralba, Elaiza. 2020 "Census Undercount of Latinos Could Cost L.A. Dearly in Funds for Critical Health and Social Services." *UCLA Newsroom*. March 24, 2020.

United States Bureau of Economic Analysis. 2012. "What is the Value of Household Work?" Bureau of Economic Research. June 11, 2012.

United States Bureau of Labor Statistics. 2015. "Labor Force Statistics from the Current Population Survey: How the Government Measures Unemployment."

United States Bureau of Labor Statistics. 2020. "News Release: The Employment Situation."

United States Bureau of the Census. 2012. "Census Bureau Releases Estimates of Undercount and Over-count in the 2010 Census."

United States Bureau of the Census. 2020. "Why We Conduct the Decennial Census." www.census.gov.

United States Bureau of the Census. 2021. "Will You Count? African Americans in the 2020 Census." Census Bureau. US Centers for Medicare and Medicare Services. Children's Health Insurance Program.

United States Department of Commerce. 1978. The Revolution in United States Government Statistics 1926–1976. Washington: U.S. Government Printing Office.

United States Department of Justice. Federal Bureau of Investigation. "What We Investigate: White-Collar Crime."

United States Department of Justice. Federal Bureau of Investigation. 2011. Uniform Crime Report. Crime in the United States, 2011.

United States Department of Justice. Federal Bureau of Investigation. 2012. "Financial Crimes Report, 2011–2012." February 27, 2012.

United States Department of Justice. Federal Bureau of Investigation. 2013. "UCR Program Adds Human Trafficking Offenses to Data Collection, Includes More Specific Prostitution Offenses." May 7, 2013.

United States. Department of Justice. Federal Bureau of Investigation. 2017. "Crime & Victimization in the United States." National Criminal Justice Reference Service.

United States Department of Justice. Federal Bureau of Investigation. 2018. Uniform Crime Reporting Program. https://www.fbi.gov/services/cjis/ucr

United States Department of Justice. Federal Bureau of Investigation. 2019. Uniform Crime Reporting Statistics. "The Nation's Two Crime Measures."

United States Department of Justice. Federal Bureau of Investigation. 2020. Internet Crime Complaint Center (IC3).

United States General Accountability Office. March 2016. Retirement Security: Shorter Life Expectancy Reduces Projected Lifetime Benefits for Low Earners. Report 16-354.

Vittert, Liberty. 2018. "Why the US Needs Better Crime Reporting Statistics." *Conversation*. October 12, 2018.

Volk, T. and J. Atkinson. 2008. "Is Child Death the Crucible of Human Evolution." *Journal of Social, Evolutionary, and Cultural Psychology*. Proceedings of the 2nd Annual Meeting of the NorthEastern Evolutionary Psychology Society.

Wadhams, Nick and Jennifer Jacobs. 2020. "China Concealed Extent of Virus Outbreak, U.S. Intelligence Says." *Bloomberg*. April 1, 2020.

Walejko, Gina et al. 2019. "Researching the Attitudes of Households Reporting Young Children — A Summary of Results from the 2020 Census Barriers, Attitudes, and Motivators Study (CBAMS) Survey." U.S. Census Bureau Final Report.

Wallace, Gregory. 2019. "Why Does the Trump Administration Want a Citizenship Question on the Census?" CNN Wire Service. July 6, 2019.

Walsh, Anthony and Cody Jorgensen. 2017. *Criminology*. Third Edition. Sage Publishers. Thousand Oaks, Ca.

Walsh, Anthony and Craig Hemmens. 2018. *Introduction to Criminology: A Text/Reader* 4th Edition. Sage Publishers. Thousand Oaks, Ca.

Wang, Hansi Lo. 2018. "Some Japanese-Americans Wrongfully Imprisoned During WWII Oppose Census Question." NPR December 26, 2018. www.npr.org/2018/12/26/636107892/some-japanese-americans-wrongfully-imprisoned-during-wwii-oppose-census-question

Wang, Hansi Lo. 2019. "2020 Census Could Lead to Worst Undercount of Black, Latino People in 30 years." NPR. June 4, 2019.

Wang, Hansi, Lo. 2021. "Trump's Census Director to Quit After Trying to Rush Out 'Indefensible' Report." NPR. January 18, 2021.

Waxman, Olivia B. 2017. "How the Public Learned About the Infamous Tuskegee Syphilis Study." Time. https://time.com/4867267/tuskegee-syphilis-study

Webster, Tony. 2020. "Counting for Dollars 2020: Estimating Fiscal Costs of a Census Undercount to States." George Washington Institute of Public Policy. May 12, 2020.

Weichselbaum, Simone and Weihua LI. 2020. "As Coronavirus Surges, Crime Declines in Some Cities." The Marshall Project, March 27, 2020.

Welna, David. 2020. "Coronavirus Has Now Killed More Americans Than Vietnam War." *NPR*. April 28, 2020.

Wenger, Jeffrey B. and Kathryn A. Edwards. 2020. "Is the Unemployment Rate Now Higher Than It Was in the Great Depression?" Commentary, The Rand Blog. May 7, 2020.

West, K. K., and Fein, D. J. (1990). "Census Undercounts: An Historical and Contemporary Sociological Issue." *Sociological Inquiry* 60 (2): 127–41.

Wexler, Chuck. 2018. "Crime Has Been Changing and Police Agencies Need to Catch Up." In *The Changing Nature of Crime and Criminal Investigations*. Police Executive Research Forum.

Wiegand, Bruce. *Off the Books: A Theory and Critique of the Underground Economy.* General Hall, Inc. Dix Hall, New York.

Williams, Chris A. and Tom Avril. 2020. "Coronavirus Death Toll May Be Undercounted by Thousands in Pennsylvania And New Jersey." *Philadelphia Inquirer.* May 26, 2020.

Williams, John. 2020. Alternative Unemployment Charts. Shadow Government Statistics (Website): Analysis Behind and Beyond Government Economic Reporting.

Wines, Michael. 2020. "It's the Official Start to the 2020 Census. But No One Counted on a Pandemic." *New York Times.* April 3, 2020.

Wolfers, Justin. 2020. "G.D.P. Doesn't Credit Social Distancing, but It Should." *New York Times.* May 14, 2020.

World Health Organization. 2014. "Large Gains in Life Expectancy." WHO News Release. May 15, 2014.

World Health Organization. 2016. "Life Expectancy Increased By 5 Years Since 2000, But Health Inequalities Persist." WHO News Release. May 19, 2016.

Woytinsky, W.S. 1941. "Controversial Aspects of Unemployment Estimates in the United States." *Review of Economic Statistics* 23: 68–77.

Yakowicz, Will. 2021. "U.S. Cannabis Sales Hit Record $17.5 Billion As Americans Consume More Marijuana Than Ever Before." *Forbes.* March 3, 2021.

Yen, Hope and Calvin Woodward. 2020. "*AP FACT CHECK: Trump Baselessly Cites Fraud in Virus Toll.*" Associated Press. October 31, 2020.

Yoon, Seokhee. 2015. "Why Do Victims Not Report?: The Influence of Police and Criminal Justice Cynicism on the Dark Figure of Crime." Doctoral Dissertation. City University of New York.

Yung, Corey Rayburn. 2014. "How to Lie with Rape Statistics: America's Hidden Rape Crisis." *Iowa Law Review* 99 (1197).

Zamarripa, Christi. 2019. "State Efforts to Support the Census." *National Council of State Legislatures* 24, no. 34 (September).

Zarroli, Jim. 2016. "Life Expectancy Study: It's Not Just What You Make, It's Where You Live." *NPR.* April 11, 2016. https://www.npr.org/sections/thetwo-way/2016/04/11/473749157/its-not-just-what-you-make-its-where-you-live-says-study-on-life-expectancy

Ziv, Shahar. 2020. "Don't Be Fooled by Official Unemployment Rate Of 14.7%; the Real Figure Is Even Scarier." Forbes. May 10, 2020.

Zurik, Lee and Jamie Grey. 2019. "Measure of Hate Documentary Explores the Undercounting of Crimes in America." NBC.12.com WVUE (Investigate TV). August 14, 2019.

Index

2020 decennial census, proposed
citizenship question, 69–71,
77, 82, 86

Abbott, Edith, 55
acquired immunodeficiency
syndrome. *See* AIDS
AFL (American Federation of Labor),
15–16, 18, 20
AIDS (acquired immunodeficiency
syndrome), 34
alternative measurements:
crime, 65
labor, 28–29, 98
American Federation of Labor. *See* AFL
American Statistical Association, 15
anti-vax movement, 35
apportionment, 70, 72–73, 80, 86
Arizona:
get out the count effort, 82
One Arizona initiative, 82
Azar, Alex, 38

Bancroft, Gertrude, 20
BEA (Bureau of Economic
Analysis), 90–91
Berkman, Lisa, 48
Bhutan, 5, 87, 89, 96–98

BJS (Bureau of Justice
Statistics), 3, 65–66
Black communities:
and census undercounts,
4, 73, 79–81
and life expectancy, 46
Tuskegee syphilis study, 35
Blanco, César, 81
BLS (Bureau of Labor Statistics), 1–2,
4, 11–31, 87
U-2 measure, 28
U-4 measure, 28
U-5 measure, 29
U-6 measure, 41
Bolsonaro, Jair, 8
Bradley, Elizabeth, 48
Bretton Woods meeting, 88
Brynjolfsson, Erik, 96
Bureau of Economic Analysis. *See* BEA
Bureau of Justice Statistics. *See* BJS
Bureau of Labor Statistics. *See* BLS
Bureau of the Census. *See*
Census Bureau
Bush, George W., 57
Byrd, James, 58

California, Complete Count
Committee, 71
Canada, 45, 93

cannabis, 61, 92–93
  and crime statistics, 61
  and GDP, 92
canvassing, 4, 15–16, 19, 22, 41, 69–70, 81, 84
cargo theft, and crime statistics, 58
Carlson, Tucker, 7
Carmona, Richard H., 46
Carter, Jimmy, 2
Case, Anne, 45
CDC (Centers for Disease Control and Prevention), 1–3, 33–51, 75
  controversies, 37
  COVID-19 fatality data, 98
  Division of Unintentional Injury Prevention, 38
  early history, 33
  Epidemic Intelligence Service, 34
  and Flint water crisis, 75–76
Census Bureau, 1, 3, 22, 79
  1940 census, 82
  Complete Count Committees, 84
  Current Population Survey, 26
  effects of undercounting, 100
  and Hate Crimes Report, 58
  and Hispanic/Latinx communities, 73, 83
  and Japanese internment, 82
  mistrust of, 78
  and NCVS, 65
  overcount of whites, 83
  and social programs, 86
  and systemic racism, 74
  undercounts of children, 76–77, 79–80
  and undocumented immigrants, 86
census data:
  accuracy of, 71–72, 76
  and affluent districts, 84
  and political representation, 73, 80, 85
  and social programs, 86
census overcounts, 5, 7, 83
census undercounts:
  Black communities, 80
  children, 76–77, 79–80
  factors influencing, 78
  and Flint water crisis, 75–76, 79
  foreign-born residents, 83
  Hispanic/Latinx communities, 81
  and slavery, 80
Center for Public Affairs Research, 84
Centers for Disease Control and Prevention. *See* CDC
charitable work, overlooked by GDP, 90–91
children, 55, 76, 79–80, 82
  and 2010 decennial census, 79
  and 2020 decennial census, 79
  census undercounts, 71, 76–77, 79–80
  crimes against, 63, 99
  and domestic violence during COVID-19 pandemic, 60
  effects of census undercounts on social programs, 79
  and Flint water crisis, 75–77
  and hard-to-count families, 76–77
  and life expectancy, 42–43, 45, 49–51
  overcount in white communities, 76, 81, 84
  racial disparities in census counts, 76, 79
  and Salk vaccine, 34
  underestimation of child abuse, 54, 60
Children's Aid Fund, 38
citizenship question for 2020 decennial census, 69–71, 77, 82, 86
Clinton, Bill, 35
CNCS (Corporation for National and Community Service), 90
Collins, Michael, 77
Complete Count Committees:
  California, 71
  Nebraska, 84
Congress, 56, 58, 88–89
conservative politics, 3, 22–23, 70, 86

Constitution, 80
Cooper, Yvette, 45
Corporation for National and
    Community Service. *See* CNCS
correctional facilities, 27
counterfeiting, 3, 57
COVID-19 pandemic, 4–9, 29, 36–37,
    54, 60, 70–71
        and 2020 census, 71, 81
        anti-vax movement, 35
        conspiracy theories, 7
        and crime reports, 54
        and GDP, 95
        misinterpretation of statistics, 6
        politicizing statistics, 5–6, 9
        reporting deaths, 4, 7–9, 37
CPS (Current Population
    Survey), 22, 26–28
Crime Index, 41, 57, 99
crime statistics, 53, 55, 57, 59–60,
    62–63, 66–68, 99
        alternative sources for
            reporting, 65
        BJS, 66
        cannabis, 61
        cargo theft, 58
        classifications, 56
        creation of UCR, 56
        criticism of UCR, 59
        cybercrime, 63–64
        dark figure of crime, 54, 60, 66
        early reporting, 55–56
        and Edith Abbott, 55
        and gun rights advocates, 53, 67
        hate crimes, 58
        inaccuracy, 53, 55, 68
        NCVS, 65, 67
        NIBRS, 66
        overemphasis on street crime, 64
        politicization of, 5, 67–68
        and UCR, 57, 61
        UCR Part I, 56, 67
        UCR Part II, 67
        undercounting, 3, 54, 59–60
        and unemployment, 11

white-collar crime, 53, 62, 64
criminal activity:
        cybercrime, 53, 55, 63–64, 68
        and GDP, 92–93
        hate crimes, 58, 65
Crofts-Pelayo, Diana, 71
Current Population Survey. *See* CPS
cybercrime, 53–55, 62–64, 67–68

dark figure of crime, 54, 60, 65–66
Darrow, Clarence, 56
Davis, James, 17–18
death:
        COVID-19, 6–8
        heroin overdoses, 40
        infant mortality, 45
        medication overdoses, 38–40
        opioid-related, 38–39
        and racial inequality, 46
        and wealth inequality, 49
Deaton, Angus, 45
decennial census:
        2010, 79
        2020, 69–72, 77, 82, 86
        and apportionment, 1
        and Black communities, 80
        and COVID-19 pandemic, 9
        and Detroit water crisis, 75–77
        and foreign-born residents, 83
        and hard-to-locate groups, 77
        history of, 13
        implications, 73–75, 85
Democratic party, 7, 24, 70
Department of Justice, 65, 93
Department of Labor, 18, 69
Detroit water crisis, 75–77, 79
deviant behavior, 55, 61–62, 67
differential counting, 80
Dillingham, Steven, 86
disease, 33–35, 45–46
        childhood, 44
        communicable, 6, 33–34
        heart, 45
        surveillance, 34
divorce, 41, 51

domestic violence, 5, 54, 59–61
Dorman, Joe, 72
drugs, 37, 45, 93
    opioid, 38–39
    prescription, 38–40
    transactions, 91, 93

economic activity:
    as basis for GDP, 5, 87,
        89–92, 94, 96
    and employment statistics, 14
    illegal, 94
    pollution-generating, 95
economy:
    and COVID-19 pandemic, 7, 9
    and crime, 63
    and GDP, 87–88, 95–96
    and illegal activities, 93
    informal, 89, 91
    and unemployment statistics, 11,
        19–20, 24, 31
education:
    coordinated, 71
    and GNH, 97
    grants, 73
    higher, 51
    for HIV prevention, 38
    and life expectancy, 51
    racial disparity, 45, 74
EIS (Epidemic Intelligence Service), 34
Electoral College, 73
embezzlement, 3, 57, 99
employers:
    and discouraged workers, 28
    and life expectancy statistics, 50
    and unemployment statistics, 16,
        21–22, 25, 27, 31
    and USES, 16
    and women, 15
    and working-class
        unemployment, 13
employment:
    and the BLS, 2, 18
    and discouraged workers, 28
    employability, 21

    and labor market conditions, 29
    and NILF, 27
    part time, 17
    scarcity, 20, 22
    and statistics for marginalized
        groups, 24–25
    and systemic racism, 74
    and unemployment statistics, 12,
        14, 17–18, 26, 32
    and unpaid family workers, 27
environmental issues, 44, 95
Epidemic Intelligence Service. *See* EIS
ethnicity:
    and census undercounts, 79
    and hate crime statistics, 57–58

Fauci, Anthony, 7–8
FBI (Federal Bureau of Investigation),
    1, 3, 53–63, 65–68, 99
    Financial Crimes Report, 62
    and mass media, 64
    Uniform Crime Report,
        53, 55, 59–60
Federal Bureau of Investigation. *See* FBI
Federal Emergency Relief
    Administration. *See* FERA
Federal Reserve, 88
fentanyl, 38–40
    See also opioids
FERA (Federal Emergency Relief
    Administration), 20
Financial Crimes Report, 62
Fitzgerald, Brenda, 37–38
Flint water crisis. *See* Detroit
    water crisis
flu, 7, 35–36
    seasonal, 6
    swine, 6, 35
    vaccines, 36
Ford, Gerald, 2, 36
Foreign Quarantine Service. *See* FQS
forgery, 3, 57
Fox News, 7
FQS (Foreign Quarantine Service), 34
fraud, 3, 57, 62–63, 99

credit, 60
  medical, 62
  stock market, 62
funding:
  allocating, 75
  misdistribution, 69
  public, 100

Gans, Herbert, 11
GAO (Governmental Accountability
  Office), 47, 70
Garrett, Laurie, 38
GDP (gross domestic product), 5, 68,
  87–89, 91–98
  Bretton Woods meeting, 88
  goods and services, 87, 94–95
  omission of charitable
    work, 90–91
  omission of criminal activities, 92
  omission of criminal
    activity, 91–93
  statistics, 93–96
gender, 48–49, 58
gerrymandering, 80
get-out-the-count initiatives, 69–73,
  82, 84, 86
Global Financial Crisis, 94
GNH (Gross National Happiness Index),
  5, 87, 96–98
GNHUSA (Gross National
  Happiness USA), 98
Golikova, Tatiana, 9
Gonsalves, Gregg, 38
goods and services, 87, 94–95
Gordon Committee, 23–25
government:
  and 2020 census, 70, 72
  and apportionment, 85
  and census counts, 41
  and census undercount of
    minorities, 74, 78, 81, 83
  and COVID-19 pandemic, 8–9
  and crime statistics, 41, 56, 67–68
  and Detroit water crisis, 77
  and early census efforts, 14

and GDP, 88, 90
and the Great Depression, 20
and hate crime statistics, 58
and life expectancy
  statistics, 40, 46
and mass media, 6
and public health measures, 44
and unemployment
  statistics, 30, 41
Governmental Accountability
  Office. *See* GAO
Great Depression, 13, 16, 18–20, 30
Great Recession, 71, 76, 95
gross domestic product. *See* GDP
Gross National Happiness
  Index. *See* GNH
Gross National Happiness USA.
  See GNHUSA

Harding, Warren G., 17
hard-to-count groups, 9, 71–72, 76–78
hate crimes, 58, 65
Hate Crimes Report, 58
Hate Crime Statistics Act, 57
Health and Human Services
  Department, 38
Heller, Jean, 35
Hendrix, Craig, 37
hierarchy rule, 59, 66
Hispanic communities, 4, 45,
  79, 81–82, 84
HIV, 37–38
homeless communities, 61, 68, 76, 78
  criminalization of, 61
homicides, 46, 59, 65
Hoover, Herbert, 18–19
Hopkins, Harry, 20
housewives, 15, 20, 24, 90
housing, 44, 74, 76, 78
Human Rights Watch, 97
human trafficking, 58–59
Hume, Britt, 7
Hurricane Katrina, 6

IC3 (Internet Fraud Complaint
    Center), 54, 63
Idaho, 84
Illinois, 72–73
illness, 11, 35–36
IMF (International Monetary Fund), 88
immigrant communities, 69, 77,
    82–83, 85–86, 99
Immigration Reform and Control
    Act. *See* IRCA
Index of Sustainable Economic
    Welfare, 96
India, 97
Industrial Employment Survey
    Bulletin, 16
industry:
    automobile, 18
    depressions, 15
    and the environment, 44
    mechanical, 16
infant mortality, 43–44, 49
informal economy, 89, 91
infrastructure, 74, 82, 95
insider trading, 62
institutionalized racism, 4, 71, 73–74,
    76–77, 79, 82, 84, 86, 98, 100
International Day of Happiness, 97
International Monetary Fund. *See* IMF
Internet Fraud Complaint
    Center. *See* IC3
internment camps, 82
Iowa, 84
IRCA (Immigration Reform and
    Control Act), 83

Japan, 44–45, 50
Japanese-American
    internment camps, 82
Japan Tobacco, 38
Jefferson, Thomas, 98

Kennedy, John Fitzgerald, 24
Kuznets, Simon, 88–90

labor force, 11–13, 20, 22, 27–29, 32,
    59, 96, 98
    civilian, 28–29
    concept of, 14, 20, 25–26, 29–31
labor force participation rate. *See* LFPR
labor market, 18, 23, 26, 29, 31
    alternative indices, 28
    conditions, 22, 28–32, 98
    fluctuations, 17
    hardship, 16, 19, 22, 29
labor statistics, 2, 12, 14–15, 17
labor supply, 20–21, 26
labor underutilization, 28–29
labor unions, 14–15
Lagarde, Christine, 96
Langmuir, Alexander, 33–34
Latinx communities, 71–73, 76, 81–83
law enforcement, 3, 54–61,
    63–64, 66–67, 99
layoffs:
    neglecting, 18
    temporary, 27
    wide-scale, 17
LEB (life expectancy at birth), 40–41
    data, 43
    and differences by nation, 50
    and gender disparities, 48
    and infant mortality, 43, 45
    vs. life span, 42
    and racial disparities, 46
    and underestimation of life
        expectancy, 42
Lebergott, Stanley, 21
legislation, 57–58, 61, 82, 92
Levine, Mark, 8
Levitan, Sar, 32
LFPR (labor force participation
    rate), 29, 31
LGBTQ+ communities, 38, 57–58, 77
    and hate crime statistics, 57, 77
life expectancy, 33–51
    average, 33, 40, 43, 47
    Black males, 46
    and cancer, 46–47
    cohort, 41

data, 40–41, 43
disparities, 46
global, 50
and illness, 35–36
and medical intervention, 43–44
Middle Ages, 44–45
and wealth, 49–51
life expectancy at birth. *See* LEB
life expectancy statistics, 1, 3, 40,
   42, 50–51, 65
lifespan, 42
   expected, 44
   vs. life expectancy, 42–43
   shortened, 45
Lily, Eli, 38
Limbaugh, Rush, 7
Long, Clarence, 30
longevity, 1, 3, 41, 43–44, 46
low-income communities, 41, 48, 76, 78

malaria, 33–34
marginalized groups, 4, 24, 65,
   71, 75, 82
marijuana. *See* cannabis
mass media, 4, 6, 29, 45, 64, 85
mass shootings, 67, 95
Matthew Shepard and James Byrd Jr.
   Hate Prevention Act, 58
MCWA (Office of Malaria Control in
   War Areas), 33
Medicaid, 73–74, 79, 82
medical intervention, 43–44
Medicare, 47, 73
Metropolitan Life Insurance
   Company, 15
Metzger, Kurt, 76
Mexico, 83
Michigan, 73, 76
military, 37–38
minority communities, 4, 69,
   77, 85, 100
   census undercounts, 74, 98
   linguistic, 77
   religious, 97
   youth, 64

misallocation of resources, 9, 61, 75
misery index, 2
MMWR (Mortality and Morbidity
   Weekly Report), 34
Mortality and Morbidity Weekly
   Report. *See* MMWR
mortality rates, 8, 45, 50
   infant, 40, 42, 45
   opioid, 40
Mountin, Joseph, 33
murders, 3, 55, 59
Muslim-Americans, 82

NAS (National Academies of Sciences,
   Engineering and Medicine), 47
National Bureau of Economic Research
   (NBER), 88
National Crime Victimization
   Survey. *See* NCVS
National Incident-Based Reporting
   System. *See* NIBRS
national income, 89, 95–96
national income accounts, 90–91, 95–96
National Industrial Conference
   Board, 15–16, 20
National Institute of Allergy and
   Infectious Diseases, 7
National Institute of Justice, 59
National Law Center on Poverty and
   Homelessness. *See* NLCPH
National Low Income Housing
   Coalition, 82
National School Lunch Program, 79, 81
National Swine Flu Immunization
   Program, 36
National Use-of-Force data, 58
Native Americans, 71, 79
NBER (National Bureau of Economic
   Research), 88
NCVS (National Crime Victimization
   Survey), 3, 41, 53–54, 65
Nebraska, Complete Count
   Committee, 84
neoliberalism, 1, 12, 14,
   30–31, 87–89, 96

newborns, 40, 42
New Jersey, 8
New York, 7, 18, 47, 72–73, 76
NIBRS (National Incident-Based
	Reporting System), 65–66
NILF (not in the labor force), 20, 22, 27
NLCPH (National Law Center on
	Poverty and Homelessness), 61
non-citizens, 70–71
non-English speakers, 76
not-in-the-labor force. *See* NILF

Obama, Barack, 47, 66
obesity, 45–46, 49
Office for National Statistics (ONS), 49
Office of Malaria Control in War
	Areas. *See* MCWA
O'Hare, William, 78
Ohio, 73
Oklahoma, 72, 79
opioids, 38–40
overdose statistics, 38–40
Owens, Candace, 7

parents, 46, 76, 84
part-time workers, 28, 31
Pence, Mike, 7
Pennsylvania, 8, 73
pensions, 41, 50, 99
Persons, Charles, 19
Pew Research Center, 58
Philip Morris International, 38
police. *See* law enforcement
political elites, 31–32
political representation, 4, 71, 73, 75,
	81, 84–86, 100
	See also apportionment
politicians, 8, 13, 20, 29, 64, 94
pollution, 95–96
poverty, and census undercounts, 84
Price, Tom, 38
property, 57, 66, 99
prostitution, 5, 57, 59, 91, 93
Proxmire, William, 35
public health:

crisis, 6, 9, 54, 60, 81, 95
	measures, 7, 44
	officials, 36, 39, 75
Public Health Service, 34–35
Public Health Watch, 49
Pursuit of Happiness Day, 98

Quarterly Household Survey, 65
Quetelet, Adolphe, 60

racial inequality, 46, 58, 79
rape, 3, 55, 57–59, 65
	See also sexual assault
Reagan, Ronald, 2
recession, 5, 24, 26, 29, 78
Redfield, Robert, 37–38
redistricting, 85
refugees, 77
religion, 58
religious, 38
renters, 71, 79
Republican party, 70
resources:
	misallocation, 9, 61, 75
	public, 69, 71
retirement, 47, 99
	age of, 1, 3, 41, 47, 50–51, 99
	decisions, 51
	delayed, 47
	early, 19
Reynolds American, 38
Ricketts, Pete, 84
robbery, 3, 55, 57, 60, 62–63
Roosevelt, Franklin Delano, 19–20
Russia, 9

Salk vaccine, 34
schools, 26, 45, 74, 84
	See also education
Senate, 17–18
sex trafficking, 59
sexual assault, 59, 93
	expanded definition, 65
sexual orientation, 57–58, 77
Shadow Government Statistics, 29

Shepard, Matthew, 58
social class:
    and crime statistics, 62, 64
    disparities, 48
    and life expectancy, 46, 49, 51
social policies, 1, 11, 14, 40, 49–50, 68
social problems, 9, 11, 14, 55
social programs, 4, 23, 75, 77, 86, 100
Social Security, 1, 3, 41, 47, 50–51, 99
    benefits, 47, 51
    and private pensions, 50
    reform, 50
social services, 73, 77
social unrest, 78
South America, 34, 83
South Dakota, 84
Southern Poverty Law Center, 58
statistical artifacts, 29, 40, 81
stereotypes, 24, 68, 99
Stiglitz, Joseph, 96
strokes, 7, 46
Sudan, 47
suicides, 45
Supplemental Nutrition Assistance
    Program, 81
Supreme Court, 86
Surgeon General, 38
Sutherland, Edwin H., 61–62
syphilis, 35
    See also Tuskegee Study
systemic racism. *See*
    institutionalized racism

tax evasion, 62
temporary jobs, 27–28
Texas, 72–73, 79
theft, 3, 55, 57, 60, 91
    cargo, 58
transportation, 17, 35, 74
Trump, Donald:
    and 2020 census citizenship
        question, 69–71, 82, 86
    and CDC, 38
    and COVID-19 pandemic, 6–8, 37
    opioid commission, 39

and undocumented
    immigrants, 86
tuberculosis, 44
Tuskegee Study, 35

U-2 measure, 28
U-4 measure, 28
U-5 measure, 29
U-6 measure, 41
UCR (Uniform Crime Report):
    alternatives to, 65
    criticisms of, 59, 67
    focus on street crime, 64
    Hate Statistics Program, 58
    human trafficking, 59
    and internet crime, 62
    and legal redefinitions, 61
    modifications to, 57
    and NCVS, 3, 53, 65
    and NIBRS, 66–67
    origin of, 56
    Part I, 3, 53
    Part II, 3, 56–57
    and rape, 58
    and street crime, 64
    Summary Reporting System, 65
    underreporting, 59–60
    and white-collar crime, 61–62
undercounts:
    and apportionment, 73, 75
    Black communities, 79, 81
    census, 41, 71–74, 76, 84–85
    children, 79
    and COVID-19 pandemic, 8–9, 29
    crime, 54, 59
    Detroit water crisis, 76–77
    foreign-born residents, 83
    Hispanic/Latinx
        communities, 82–83
    minority groups, 4, 74, 100
    unemployment, 31
underemployment, 22, 32
undocumented communities, 72–73, 82
unemployment, 2, 4–5, 11–31, 69, 98
    data collection, 13, 15, 17, 19, 41

long-term, 2, 11
measuring, 14–15, 17, 20,
  22–23, 25–26
rates, 1–2, 12, 14, 18,
  22–23, 31–32
tolerable rates of, 31
unemployment statistics, 2, 12
  alternative measures, 28, 65
  and CPS, 26
  criticism of, 28
  evolution, 13, 16, 24
  official, 22, 24, 30, 98
  politicization of, 25, 30–32, 99
Uniform Crime Report. *See* UCR
United Nations General Assembly, 97
United States Employment Service, 18
University of Maryland's Genuine
  Progress Indicator, 96
USA Patriot Improvement and
  Reauthorization Act of 2005, 58

vaccines, 34–37, 44
  HIV, 37
  influenza, 34

wages, 13, 16, 27, 44
Wagner, Robert, 17
Wangchuck, Jigme Singye, 97
wealth, and life expectancy, 46, 48
weapons, 57, 92
Webb, John, 21
welfare programs, 12, 14, 24, 44, 51,
  89, 96, 98
well-being, 5, 68, 87–88, 95–97
  collective, 98
  initiatives, 98
  national, 5, 89
  psychological, 97
  social, 89, 97
West Africa, 34, 63
West Virginia, 79, 84
white Americans, 45
white-collar crime, 3, 53, 61–62,
  64, 68, 99
white communities, 44–46, 74, 81, 84

and overcounting, 84–85
  working-class, 45
White House Coronavirus Task Force, 7
White House Office of National Drug
  Control Policy, 39
white nationalism, 58
WHO (World Health
  Organization), 36, 49–50
William Wilberforce Trafficking Victims
  Protection Reauthorization Act, 58
women:
  Black, 48
  and employment, 49
  and wealth, 48–49
  white, 45, 48
women. and children, 55, 60, 99
workers, 91
  and COVID-19 pandemic, 9
  demand for, 21
  discouraged, 2, 14, 28–29
  displaced, 11
  employability, 21
  employed, 17–18, 27
  and GDP, 88, 96
  idle, 17
  and job satisfaction, 11
  and layoffs, 19
  NILF, 27
  and pensions, 50
  retired, 26
  salary, 27
  and strikes, 25
  temporary, 26–27
  undocumented, 69
  unemployed, 13–14, 16, 23, 31
  and unemployment statistics,
    12, 17, 31
  unpaid, 26
  unskilled, 21
workplace conditions, 12
Works Progress
  Administration. *See* WPA
World Bank, 88
World Economic Forum, 96
World Health Organization. *See* WHO

World War I, 15
World War II, 22, 33, 82
WPA (Works Progress Administration),
  20–22, 25–26

Wright, Carroll, 14
WTO (World Trade Organization), 88
Wyoming, 84

About the Author

Robert E. Parker is full professor of sociology at the University of Nevada, Las Vegas. He is the author of *Flesh Peddlers and Warm Bodies: The Temporary Help Industry and Its Workers*, and coauthor of *Building American Cities: The Urban Real Estate Game*.